Building SPAs with Django and HTML Over the Wire

Learn to build real-time single page applications with Python

Andros Fenollosa

BIRMINGHAM—MUMBAI

Building SPAs with Django and HTML Over the Wire

Copyright © 2022 Packt Publishing

Group Product Manager: Pavan Ramchandani
Publishing Product Manager: Bhavya Rao
Senior Editor: Mark D'Souza
Content Development Editor: Feza Shaikh
Technical Editor: Simran Udasi
Copy Editor: Safis Editing
Project Coordinator: Manthan Patel
Proofreader: Safis Editing
Indexer: Tejal Daruwale Soni
Production Designer: Joshua Misquitta
Marketing Coordinators: Anamika Singh and Marylou De Mello

First published: August 2022
Production reference: 1290722

Published by Packt Publishing Ltd.
Livery Place
35 Livery Street
Birmingham
B3 2PB, UK.

ISBN 978-1-80324-019-0

www.packt.com

I want to dedicate this book to four people who gave me the necessary tools to make writing it possible:

My mother, for giving me a heart

My grandmother, for giving me an education

My uncle, Toni Hurtado, for giving me wisdom

My girlfriend, Valentina Rubane, for giving me motivation

– Andros Fenollosa

Foreword

This book is about one principle: coordination. How do we coordinate events that occur in different places and at different times within a web application? For example, if multiple users are connected to the same website and need to send messages to each other, how can we achieve fast and effective communication? Such systems are known as real-time applications.

Real-time applications are a standard in web development today. The WebSocket protocol emerged in 2011; it allows you to open a bidirectional tunnel between your web application and another component. The WebSocket protocol creates greater dynamism, and thus improves the user experience. In addition to lower resource consumption, information is updated only when necessary - this avoids redundant requests to the server.

There are many sorts of real-time applications. The first ones that come to mind are social networks or messaging applications since they are part of our daily life. We like to receive information instantly, because we don't need to refresh the page to know if an event has occurred. We don't want to wait, we are inherently impatient. Learning how to develop real-time applications gives us an advantage when finding new techniques to start a software project.

This work presents an alternative to JavaScript for creating Single-Page Applications; and some might worry that the author has started a crusade against JavaScript - but it's not the case. Instead, his interest is in explaining the power of Python using WebSockets over HTML to produce real-time applications.

The author of this book, Andros Fenollosa, is a passionate web developer who, using his extensive teaching experience, will give you the necessary foundations to be involved in the development of real-time application.

Camilo Chacón Sartori,

Ph.D. Student in Computer Science at Artificial Intelligence Research Centre (IIIA)

Contributors

About the author

Andros Fenollosa have been teaching web technologies, best practices, and the fundamentals of functional programming for more than a decade.

Since finishing his studies, he have not stopped developing applications, sometimes as a frontend developer, others as a mobile application developer, backend developer, and project manager. Currently, he is working as CTO in a software development studio.

He combine his professional work maintaining several open source projects related to Python, Django, Clojure (including my own framework: Tadam Web Framework), and Linux. He also like to share his knowledge through podcasting, writing newsletters and books, Twitter, and his blog.

About the reviewer

Norbert Máté is a senior web developer/tech lead with 13+ years of experience. His primary programming languages are Python and JavaScript. He is passionate about software architecture, clean code, and leadership. Norbert has also reviewed other Django books, such as *Django RESTful Web Services* and *Django 2 by Example*.

I would like to thank my wife for her support.

Table of Contents

Part 2: WebSockets in Django

3

Adding WebSockets to Django

4

Working with the Database

5

Separating Communication in Rooms

Part 3: HTML over WebSockets

6

Creating SPAs on the Backends

7

Creating a Real-Time Blog Using Only Django

Part 4: Simplifying the frontend with Stimulus

8

Simplifying the Frontend

Preface

HTML over WebSockets simplifies the creation of **single-page applications** (**SPAs**) by avoiding frontend rendering, giving real-time responses, and simplifying the logic by moving it to the backend. In this book, you won't learn how to use a JavaScript rendering framework such as React, Vue, or Angular, instead moving the logic to Python. This will simplify your developments, giving you real-time results with all the tools provided by Django.

Developers will learn state-of-the-art WebSockets techniques to realize real-time applications with minimal reliance on JavaScript. They will also learn, from the ground up, how to create a project with Docker, test it, and deploy it on a server.

You will follow the roadmap of a Django developer who wants to create SPAs with a good experience. You will create a project and add Docker, development libraries, Django channels, and bidirectional communication, and then you will create real projects of all kinds using HTML over WebSockets, such as a chat app or a blog with real-time comments. You will modernize your development techniques by moving from using an SSR model to creating web pages using HTML over WebSockets. With Django, you will create SPAs with professional real-time projects where the logic will be in Python.

By the end of the book, you will be able to make real-time applications and will have mastered WebSockets with Django.

Who this book is for

This book is for intermediate-level Django developers and Python web developers who wish to create real-time websites with HTML and Django channels. If you are a developer looking for an alternative to creating an API without giving up the possibility of separating the frontend from the backend and do not want to depend on JavaScript to make SPAs, then this book is for you. Basic knowledge of HTML and Python along with an understanding of basic web development concepts is expected.

What this book covers

Chapter 1, Setting up the Virtual Environment, is where we prepare the IDE and Docker to be able to work with Python.

Chapter 2, Creating a Django Project around Docker, is where we create a project in Django and test it with different Docker containers.

Chapter 3, Adding WebSockets to Django, is where we integrate a WebSockets server into Django via Channels and then test that it can send and receive data.

Chapter 4, Working with the Database, is where we create a microblogging application where we interact with a database performing elementary actions, such as querying, filtering, updating, or deleting.

Chapter 5, Separating Communication in Rooms, is where we learn how to separate communication by users or groups by creating a chat application with private messages and groups.

Chapter 6, Creating SPAs on the Backends, is where we will integrate solutions for the typical problems you may encounter when building a SPA, such as integrating a server-side rendering system per route or hydration sections.

Chapter 7, Creating a Real-Time Blog Using Only Django, helps you use all the knowledge acquired to build a complete SPA blog in real time.

Chapter 8, Simplifying the Frontend, covers simplifying the interaction with client events by integrating Stimulus into the frontend.

To get the most out of this book

You will need a version of Docker and Docker Compose (or Docker Desktop, which includes everything) installed on your computer in the latest version and a code editor or IDE. I recommend PyCharm Professional. All code samples have been tested using Docker on macOS and Linux operating systems; however, they should also work without a problem on Windows.

Software/hardware covered in the book	Operating system requirements
Docker	Windows, macOS, or Linux
Docker Compose	Windows, macOS, or Linux
PyCharm Profesional	Windows, macOS, or Linux

You will need the professional version of PyCharm to have Docker integration. Otherwise, you can start Docker from the terminal or with Docker Desktop.

If you are using the digital version of this book, we advise you to type the code yourself or access the code from the book's GitHub repository (a link is available in the next section). Doing so will help you avoid any potential errors related to the copying and pasting of code.

Download the example code files

You can download the example code files for this book from GitHub at https://github.com/PacktPublishing/Building-SPAs-with-Django-and-HTML-Over-the-Wire. If there's an update to the code, it will be updated in the GitHub repository.

We also have other code bundles from our rich catalog of books and videos available at https://github.com/PacktPublishing/. Check them out!

Download the color images

We also provide a PDF file that has color images of the screenshots and diagrams used in this book. You can download it here: `https://packt.link/2q526`.

Conventions used

There are a number of text conventions used throughout this book.

`Code in text`: Indicates code words in text, database table names, folder names, filenames, file extensions, pathnames, dummy URLs, user input, and Twitter handles. Here is an example: "The information from the frontend goes to `receive_json`, which in turn receives the `'text in capital letters'` action by executing the `self.send_uppercase(data)` function."

A block of code is set as follows:

```
* {
    font-family: "Helvetica Neue", Helvetica, Arial, sans-
serif;
    box-sizing: border-box;
}
```

Bold: Indicates a new term, an important word, or words that you see onscreen. For instance, words in menus or dialog boxes appear in **bold**. Here is an example: "At the moment, the **Edit** and **Delete** buttons are for decoration; later, we will give them their functionality."

> **Tips or important notes**
> Appear like this.

Get in touch

Feedback from our readers is always welcome.

General feedback: If you have questions about any aspect of this book, email us at `customercare@packtpub.com` and mention the book title in the subject of your message.

Errata: Although we have taken every care to ensure the accuracy of our content, mistakes do happen. If you have found a mistake in this book, we would be grateful if you would report this to us. Please visit `www.packtpub.com/support/errata` and fill in the form.

Piracy: If you come across any illegal copies of our works in any form on the internet, we would be grateful if you would provide us with the location address or website name. Please contact us at copyright@packt.com with a link to the material.

If you are interested in becoming an author: If there is a topic that you have expertise in and you are interested in either writing or contributing to a book, please visit authors.packtpub.com.

Share Your Thoughts

Once you've read, we'd love to hear your thoughts! Scan the QR code below to go straight to the Amazon review page for this book and share your feedback.

https://packt.link/r/1803240199

Your review is important to us and the tech community and will help us make sure we're delivering excellent quality content.

Part 1: Getting Started with Python

In this part, we will learn how to configure and build a Django project using Docker containers. We start with preparing a workspace, configuring a Python image, creating a Django project and integrating it so that it is isolated to the operating system, and can be lifted from any computer.

In this part, we cover the following chapters:

1

Setting up the Virtual Environment

A good programmer is not afraid of technology because their confidence doesn't lie in the programming language, but in their own skills and experience. Tools only make them more productive. We can't build even the simplest website in an acceptable amount of time without the right software. Building websites with Python is possible on any modern operating system, regardless of the hardware behind it. The core team that maintains this fantastic language already takes care of some of the more tedious tasks, such as compiling it and optimizing it for the processor you're using.

However, building a web application in Python, even if we only respond with plain text, requires a great deal of knowledge, including of servers and web applications as well as the WSGI or ASGI interface. We need to abstract that complexity to respond to requests, environments, asynchrony, WebSocket, database connections, and the other elements that define a current web application. That's why we're going to set up a desktop with everything you need to be a productive modern Django developer. We will build different real-time applications using the technology offered by Channels, a Django extension (developed by the same Django team), which includes a WebSocket server and WebSocket integrations. The architecture of the applications will differ from how server-side rendering works. The communication path between the server and the client will be bidirectional, allowing us to use it to receive or send events and/or HTML. My intention is that upon finishing the chapter, your focus will be on the code and not on complex configurations that may distract you. To achieve this, we will make use of Docker, the famous container manager, which will open up the possibility of adding all kinds of software already precooked to launch without investing practically any time: databases, web servers, mail servers, and caches, among others. Don't worry if you have no experience with Docker. I'll teach you the basics without going into low-level details. After a few tweaks, you'll practically forget that it's running in the background.

It's important not only that we know how to write Python and create real-time infrastructures with Django but also that we have the skills to be independent of the operating system when deploying or working in a team. By virtualizing (or isolating) the processes, we can remain unconcerned about the operating system where it runs, making the project easy to continue for any specialist, and we can anticipate future problems that may occur when deploying to a production server.

In this chapter, we'll be covering the following topics:

- Exploring the software required
- Adding dependencies
- Configuring the IDE
- Installing Django
- Creating our project

Exploring the software required

In this section, we will take a look at the software that we'll be using throughout the book and how to install it. The code for this chapter can be found at `https://github.com/PacktPublishing/Building-SPAs-with-Django-and-HTML-Over-the-Wire/tree/main/chapter-1`.

Operating system

You should work on an operating system that supports Docker, such as one of the following:

- Linux distribution, preferably **Ubuntu** or **Debian**
- macOS in its latest version
- Windows 10 or higher, preferably with the Linux subsystem active and Ubuntu or Debian installed
- BSD descendants, preferably FreeBSD

Code editor

I assume that if you are reading this book, you already have experience with Python and you have an IDE or rich editor that is ready. If you need to change the IDE, I have recommended, from most to least highly recommended, in the following list some that I consider perfect for working with Python:

- **PyCharm Professional**: If you are a student at a recognized school, you can claim a free student license from JetBrains. Otherwise, I encourage you to pay for the license or use their demo. There is a free version of the IDE, **PyCharm Community Edition**, but you will not be able to use the Docker interpreter, as this is a feature of the Professional version. You can download this editor from `https://www.jetbrains.com/pycharm/`.

- **Visual Studio Code (VSCode)**: This is a very popular editor in web development, created and maintained by Microsoft. You can download this editor from `https://code.visualstudio.com/`.

- **Emacs**: This is very easy to use with a preconfigured framework such as Spacemacs or Doom. You can download this editor from `https://www.gnu.org/software/emacs/`.

- **Sublime Text** with the **Djaneiro** package: This is the easiest option if you are not looking for complications. You can download this editor from `https://www.sublimetext.com/`.

Don't force yourself to change. A code editor is a very personal thing, like choosing a brand of underwear: once you find one that fits your way of being, you don't want to change. I understand that you may not feel like learning new shortcuts or workflows either. Otherwise, if you have no preference, you are free to visit the website of any of the preceding editors to download and install it on your computer.

All the examples, activities, and snippets in the book will work with whatever your editor or IDE of choice is. They will mainly help you with syntax errors, autocompletion, and hints, but your code will be self-contained since it is always stored in plain text. A Python programmer is a Python programmer in any editor but not all editors work well with Python.

Python

You don't need to install it. You're reading correctly; the editor didn't make a mistake in the review. We'll use Docker to install a Python container capable of launching basic commands in Django, such as creating a project or an app or launching the development server.

I assume that if you are here, it is because you feel comfortable programming with Python. If not, I would recommend you read some of Packt's books:

- *Learn Python Programming – Third Edition, Fabrizio Romano and Heinrich Kruger, Packt Publishing* (`https://bit.ly/3yikXfg`)

- *Expert Python Programming – Fourth Edition, Michał Jaworski and Tarek Ziadé, Packt Publishing* (`https://bit.ly/3pUi9kZ`)

Docker

The fastest way to install Docker is through **Docker Desktop.**. It's available on Windows, macOS, and Linux (in beta as I write this). Just go to the official website, download, and install:

`https://www.docker.com/get-started`

In the case that you want to install it directly through the terminal, you will need to search for *Docker Engine* (`https://docs.docker.com/engine/`). This is highly recommended if you use Linux or BSD.

Also install **Docker Compose**, which will simplify the declaration and management of images and services:

`https://docs.docker.com/compose/install/`

Git

There is no development that does not involve a versioning system. Git is the most popular option and is almost mandatory to learn.

If you have no knowledge or relatively basic experience with it, I recommend looking at another of Packt's books, such as *Git Essentials – Second Edition, Ferdinando Santacroce, Packt Publishing* (`https://bit.ly/3rYVvKL`).

Alternatively, you can opt to review the more extensive documentation from the official Git website:

`https://git-scm.com/`

Browser

We will avoid focusing on the visual aspect of the browser, which means frontend implementation features such as CSS compatibility or JavaScript features do not matter. The most important thing is to feel comfortable when debugging the backend. Most of the time, we will be in the console checking that the requests (`GET`, `POST`, and the like) work as expected, watching the communication over **WebSocket** to make it smooth, and sporadically manipulating the rendered HTML.

> WebSocket
>
> WebSocket is a bidirectional communication protocol, different from HTTP, which facilitates the sending of data in real time between a server and a client, in our case, between a Django server and a frontend client.

In this book, I will use the **Firefox Developer Edition** (`https://www.mozilla.org/en-US/firefox/developer/`) browser because it is so convenient to manage the aspects mentioned using it. You are free to use any other browser, such as **Chrome**, **Safari**, or **Edge**, but I'm not sure whether all the features I will use are available with those browsers.

With the software installed, we can start working with the preparations around Python and Docker to run Django or future Python code.

Adding dependencies

We're going to run Python via Docker and a configuration file. That way, any developer can replicate our code regardless of whether they have Python installed on their machine and they will be able to run the associated services with just one command.

First, we will create a Python file called `hello.py` with the following content:

```python
print("Wow, I have run in a Docker container!")
```

The code is ready. We can continue.

The goal will be to configure Docker to run the file. Sounds easy, doesn't it? Here we go!

We'll create a file called `Dockerfile` with the following code:

```
# Image
FROM python:3.10

# Display the Python output through the terminal
ENV PYTHONUNBUFFERED: 1

# Set work directory
WORKDIR /usr/src/app

# Add Python dependencies
## Update pip
RUN pip install --upgrade pip
## Copy requirements
COPY requirements.txt ./requirements.txt
## Install requirements
RUN pip3 install -r requirements.txt
```

This file is used to create a Docker image, or template, with instructions that will be cached. Since they are precached, their launch will be almost instantaneous. Let's check out what's going on in the code:

- With `FROM python:3.10`, we are using another existing image as a base. We are extending the work already done. But... where is it? Docker has a repository full of images, or templates, called **Docker Hub** (`https://hub.docker.com/`), a place where developers selflessly upload their work. There is an official image called `python` and we mark it with a tag to use version 3.10. If you have worked with Docker before, you might be wondering why we are not using the *Alpine* version, the famous operating system that saves so much space in servers around the world. For two reasons: Python is slower (`https://pythonspeed.com/articles/alpine-docker-python/`) and it doesn't have the ability to compile dependencies. The *Slim* version also exacerbates the last problem and is recommended only for production versions that are short of space.

- `ENV PYTHONUNBUFFERED: 1` shows us the Python messages, for example, when we use `print()`. If it was not added, they would go directly to the Docker log.

- By adding `WORKDIR /usr/src/app`, we define the path in which the commands will be executed inside the Docker container, not inside our operating system. This is equivalent to changing the directory with `cd`.

- We will also take the opportunity to install the Python dependencies that we will be adding in the future inside `requirements.txt`. We update *pip* with `RUN pip install --upgrade pip`, copy the list of dependencies from the folder to the image with `COPY requirements.txt ./requirements.Txt`, and finally, run *pip* to install everything with `RUN pip3 install -r requirements.txt`.

- At the root of the project, we create another file called `docker-compose.yaml` with the following content:

```
version: '3.8'

services:

  python:
    build:
      context: ./
      dockerfile: ./Dockerfile
    entrypoint: python3 hello.py
    volumes:
      - .:/usr/src/app/
```

This is the orchestrator, a file where we define each service and its configurations. In this case, we are only going to have a service called `python`. With `build`, we tell Docker to use the image that we just defined in the `Dockerfile`. With `entrypoint`, we indicate what it should do when the service is launched: `python3 hello.py`. Finally, in `volumes`, we tell it to mount the root directory, represented by a single dot, with `/usr/src/app/`, which is an internal directory of the image. This way, the service will have access to all the files in the folder.

- Next, we create an empty file called `requirements.txt`. We will not add a single line, though the file must be present.

We're ready to go! Open the terminal, go to the working folder, and tell `docker-compose` to pull up the services:

```
cd [your folder]
docker-compose up
```

Docker will gradually perform several tasks: it will download the base `python` image, build its own image by executing the instructions we have defined, and raise the `python` service. It will print 2 lines per console, as in the following:

```
python_1 | Wow, I have run in a Docker container!
python_1 exited with code 0
```

We've executed the Python file! Victory!

With the recent ability to run Python using Docker, it's time to integrate it into an IDE to make it easier to run without using the terminal.

Configuring the IDE

PyCharm is highly popular because it's a tool specially prepared to work with Python and it also includes interesting integrations with databases, Git, HTTP clients, environments, and the like. One of the most used is certainly the one related to Docker, so I will use this fantastic IDE in future examples. However, as I said before, it is not mandatory to use it; there are enough alternatives to please everyone. All code and activities shown in this chapter will work independently of the editor.

To set up the IDE, follow these steps:

1. Open the folder where you want to work using PyCharm (**File | Open**). A directory tree will be displayed on the left.

2. Click on the Python file (`hello.py`). It's not possible to run the Python code if you don't use the terminal; PyCharm doesn't know where the Python interpreter, or executable, is otherwise. It's inside a Docker image that the operating system can't access, for now.

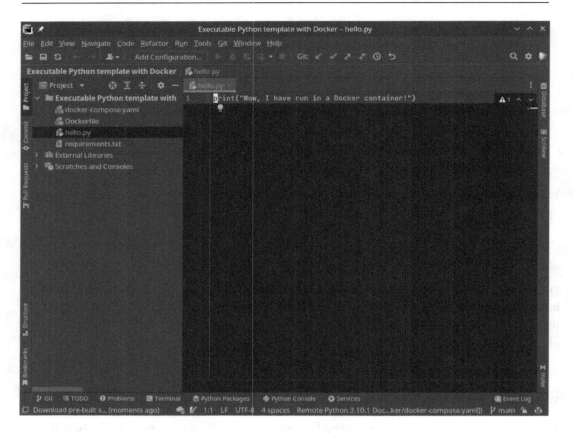

Figure 1.1 – Open the Python file

3.   PyCharm may open a popup that suggests creating a virtual environment. You can skip this step or close the window; we will use Docker for the job. If you don't see the window, you can continue without fear.

4.   We then check whether we have the Docker plugin installed. Go to **File** | **Settings** | **Plugins** and look for **Docker**.

5. If it is installed, it will appear in the **Installed** tab. If not, you will have to look for it in **Marketplace** and then click on the **Install** button. Then, restart PyCharm. Ensure you do this. Otherwise, we won't be able to continue.

Figure 1.2 – Installing the Docker plugin

6. Now open **File | Settings | Build, Execution, Deployment | Docker** and press the + button. Then, search for **Docker**.

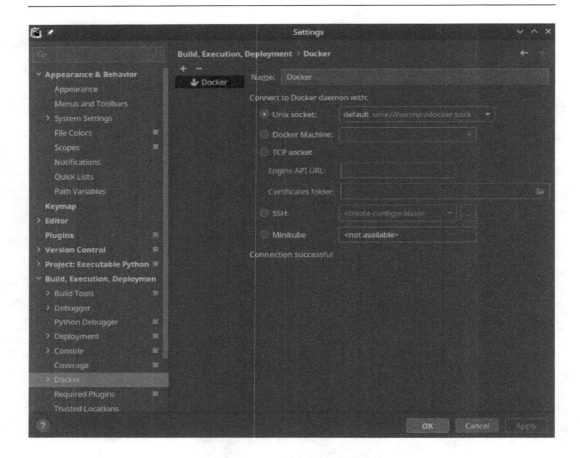

Figure 1.3 – Connecting with Docker

7. Enter `Docker` in the **Name** field, for example, and activate **Unix socket**. At the bottom, you will see the **Connection successful** message.

8. We only need to tell PyCharm not to look for the Python interpreter, or executable, on the machine (if there is one) and to use the Docker service we have created instead.

9. Go to **File** | **Settings** | **Project: Executable Python**. Here, we deploy the **Python Interpreter**, select **Remote Python xxx Docker Compose**, and click on **Apply**. The interpreter name may change depending on the folder name.

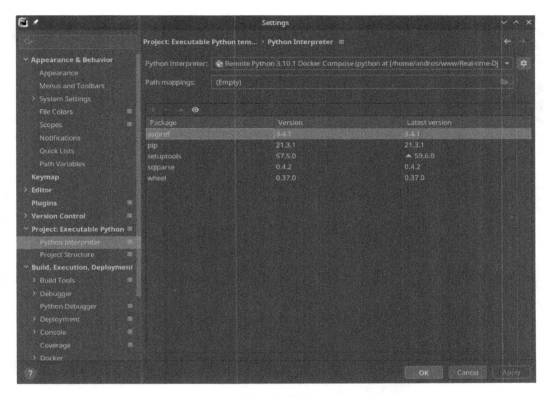

Figure 1.4 – Adding the Python interpreter

10. It will automatically detect the dependencies that are already installed on the machine but we will ignore them. By the way, this is a good place to manage all Python dependencies visually.

11. Now, it's time to run the Python code using the configuration you just made.

12. Close **Settings** and focus on the directory tree (on the left side). Right-click on `hello.py` and then **Run 'hello'**.

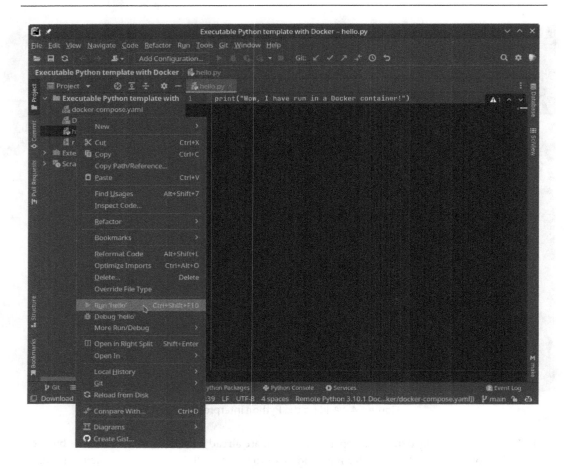

Figure 1.5 – Running Python with PyCharm

13. At the bottom of the editor, an area with the log of the execution will open. As proof that it has been done successfully, we can see the print statement.

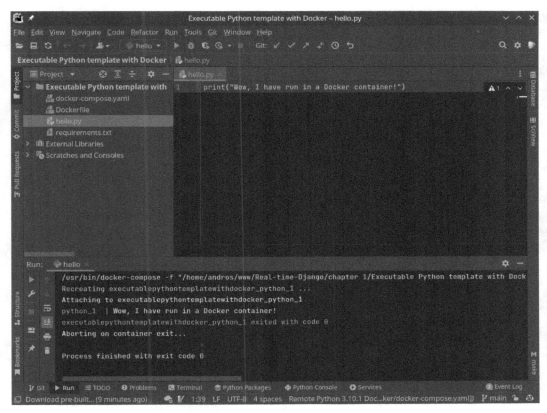

Figure 1.6 – Viewing the Python execution log through Docker integration

14. Also, if we open the `docker-compose.yaml` file, we can run the containers individually.

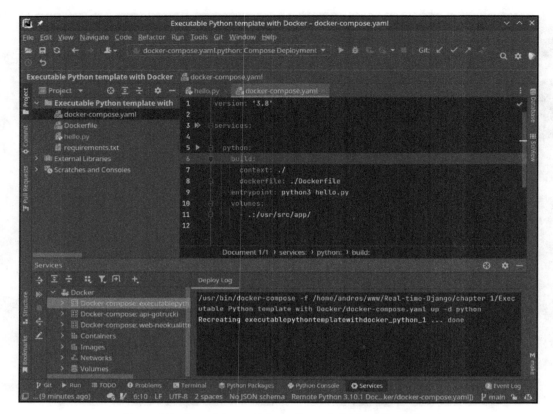

Figure 1.7 – Launching containers through Docker integration

15. On line 5 in *Figure 1.7*, you can see a green arrow; when you click on it, it will launch the service and, again, the Python code.

PyCharm is already integrated with Docker and is able to launch Python with its dependencies in isolation from the operating system. We are ready to work directly with Django. We are going to create a simple project using the official Django client to have a minimal structure when developing.

Installing Django

We already have a base with which to work with Python; now, it's time to install the minimum dependencies and tools that will be practical in Django.

We will add the following content to `requirements.txt`, which is currently empty:

```
# Django
django===4.0
```

```
# Django Server
daphne===3.0.2
asgiref===3.4.1
# Manipulate images
Pillow===8.2.0
# Kit utilities
django-extensions===3.1.3
# PostgreSQL driver
psycopg2===2.9.1
# Django Channels
channels===3.0.4
# Redis Layer
channels_redis===3.2.0
```

You may not know some of them since they are part of the project that adds WebSocket to Django. Let's review each one:

- **Django**: This automates many important tasks, such as database connections, migrations, HTML rendering, sessions, and forms. In addition, being one of the most used and active frameworks, it gives us a high degree of security.

- **Daphne**: An asynchronous server maintained by the Django team itself. We'll need it to work with WebSocket, to emit or receive data without blocking the app.

- **asgiref**: An ASGI library that needs Channels to work.

- **Pillow**: The mandatory Django library for manipulating images.

- **django-extensions**: A set of extensions that adds elements, such as *jobs*, script execution, database synchronization, and static storage in *S3*.

- **Psycopg2**: The driver to connect to PostgreSQL, the database that we will use and is most recommended to use with Django.

- **Channels**: Adds protocols and functionality for working with WebSocket to the heart of Django.

- **channels_redis**: We must have a record of the connections that we have active and the groups to which they belong. Using a database that writes to the hard disk is an inefficient way to manage it. To solve this, we'll connect with a Redis service later, as it works on volatile memory and is incredibly fast.

PyCharm may suggest you install a plugin, as shown in the following screenshot:

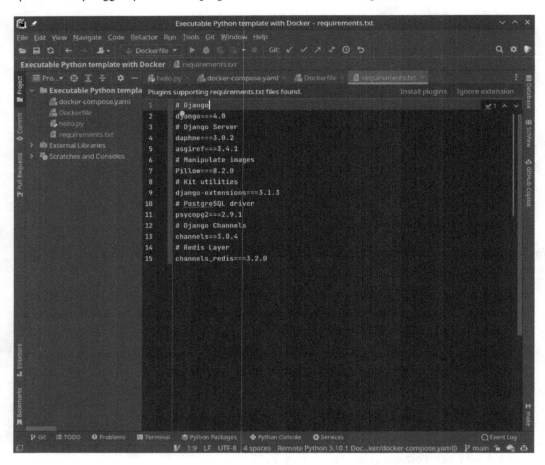

Figure 1.8 – PyCharm asking whether you want to install the new dependencies

If you click on **Install plugins**, it will show you a window, like so:

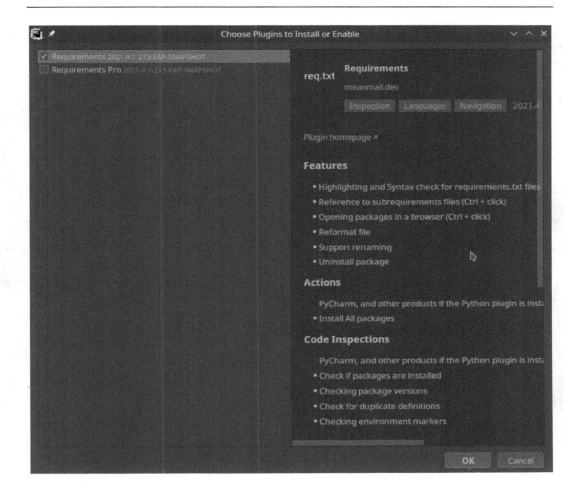

Figure 1.9 – PyCharm asking whether you want to install the requirements plugin

By clicking on the **OK** button, we can enjoy color codes for `requirements.txt`.

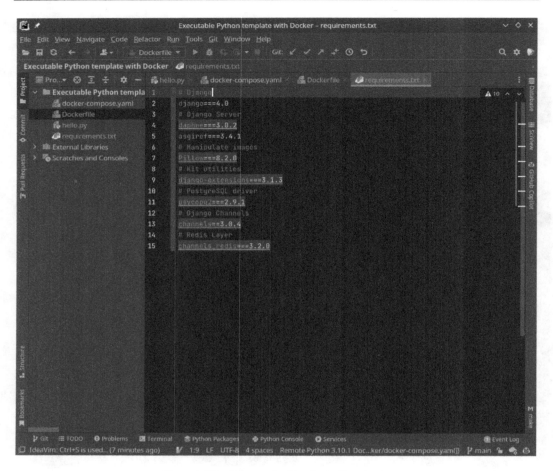

Figure 1.10 – Color codes thanks to the plugin

Now, we will recompile the image so that all the dependencies we have added are installed.

With PyCharm, this can be done in a visual way. Go to **Dockerfile**, right-click on the double arrow shown in the following screenshot, and select **Build Image for 'Dockerfile'**:

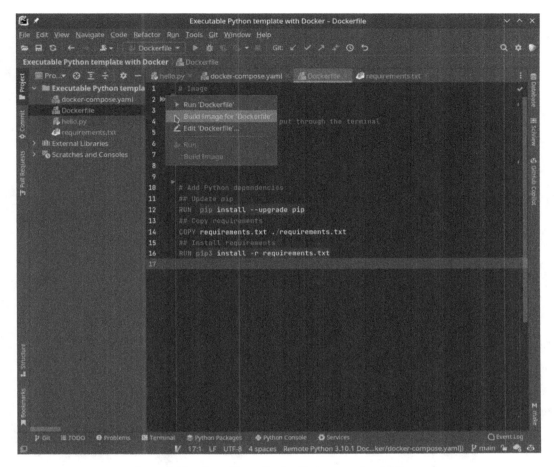

Figure 1.11 – Compiling a Dockerfile image using PyCharm

If you are using the terminal or another editor, we will use `docker-compose` in the directory:

```
docker-compose build
```

By recreating the image, we've integrated all the dependencies inside the image; now, Django has everything we need. To check that it's installed and we have version 4, we'll temporarily modify `entrypoint`:

```
Entrypoint: django-admin --version
```

And then, we'll run the service.

Remember that you can do this by clicking on the green arrow next to Python (line 5 in *Figure 1.12*) or through docker-compose.

```
docker-compose up
```

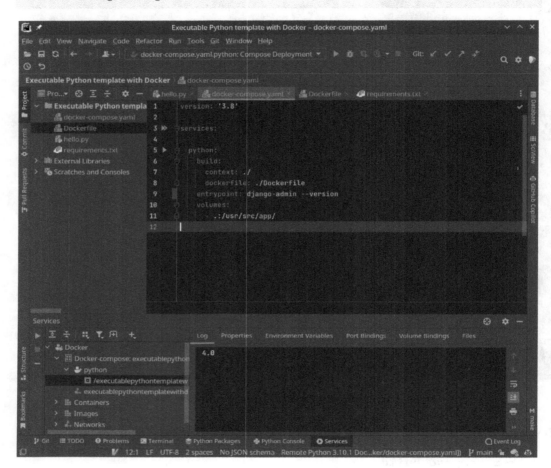

Figure 1.12 – Checking which version of Django is installed

In both cases, you can see that it returns 4.0 or the version specified in requirements.txt. We are ready!

All this work can serve as a template for future Python developments. Don't lose it!

After creating a minimal template through the Django client, we're going to configure it to launch the test server every time the service is up.

Creating our project

Django needs its own directory and file structure to work. That's why we need to generate a project via `django-admin`, a terminal client built to launch Django tasks. Don't worry! You don't have to install anything new; it was added when we added the Django dependency.

Let's build a file with shell instructions to perform all the tasks in one go. We create a file called `start-project.sh`, where we are working with the following content:

```
# Create the 'hello-word' project
django-admin startproject hello_world
# Create a folder to host the future App with the name
    'simple-app'.
mkdir -p app/simple_app
# Create the 'simple-app' App
django-admin startapp simple_app app/simple_app
```

Here is what we are doing:

- With the first instruction, `django-admin startproject hello_world .`, we're creating a project (`startproject`) called `hello_world` and, with the final dot, we're telling it to make it in the directory where we're running it.

- When we launch `mkdir -p app/simple_app`, we create a directory called `simple_app` which is inside app. The goal is to organize the apps, saving them all in the same directory; we also create the folder in which the first app will be saved: `simple_app`.

- Finally, we create the app with `django-admin startapp simple_app app/simple_app`. The `simple_app` and `app/simple_app` parameters define the app's name and its location, respectively, which we created with the previous command.

- In short, we'll call the project `hello_world`, and inside it, we'll have a single app with the original name `simple_app`.

PyCharm may suggest that you install a plugin to check for syntax problems; it's a good idea to do so.

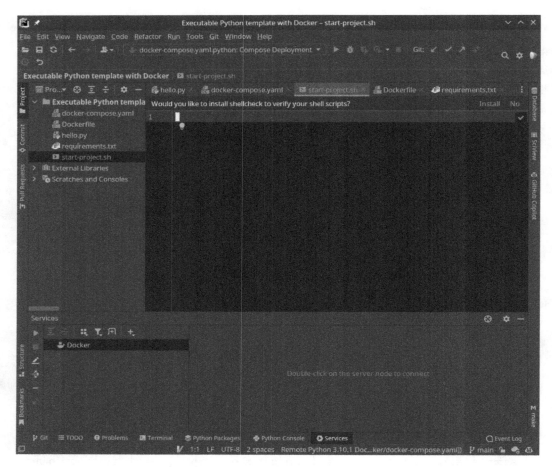

Figure 1.13 – PyCharm suggests installing a syntax checker for shell files

To execute the script, we again must temporarily modify `entrypoint` with `bash start-project.sh`:

```
version: '3.8'

services:

  python:
    build:
      context: ./
      dockerfile: ./Dockerfile
```

```
        entrypoint: bash start-project.sh
    volumes:
      - .:/usr/src/app/
```

We launch the container as we have already learned to: open the `docker-compose.yaml` file and click on the double green arrow in `services` or the single arrow in `python`.

If you are using the terminal or another editor, we will use `docker-compose` in the directory:

`docker-compose up`

When Docker finishes, the new files and directories will appear. Be patient if you don't see them in PyCharm; sometimes it has a hard time refreshing when new files appear. You can wait or right-click on any file and click **Reload from Disk**.

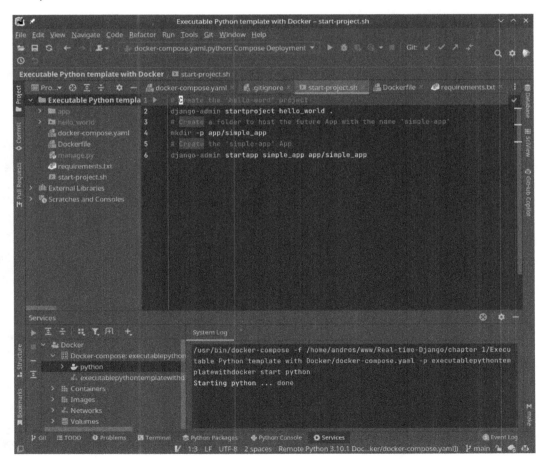

Figure 1.14 – The newly generated Django project

It's time to modify entrypoint one last time. Let's get the development server up. It's time to reap the fruits of our labor.

Modify it by adding the following:

```
python3 manage.py runserver 0.0.0.0:8000
```

If you haven't worked with Django before, manage.py is equivalent to using django-admin. The advantage of the former is that it uses the project's configuration, while django-admin is more general and you have to tell it where the configuration is; so, it's more practical to use manage.py as soon as the project exists.

The action we want to launch is to raise a development server with runserver. The 0.0.0.0:8000 parameter indicates that we are open to any *IP* that makes the request and finally, we will use port 8000 to accept connections.

On the other hand, for Docker to route port 8000 from the service to the outside, we will add ports 8000:8000 somewhere inside the service.

Altogether, it will look like this:

```
version: '3.8'

services:

  python:
    build:
      context: ./
      dockerfile: ./Dockerfile
    entrypoint: python3 manage.py runserver 0.0.0.0:8000
    ports:
      - "8000:8000"
    volumes:
      - .:/usr/src/app/
```

We launch the service again. Now, open your favorite browser and enter `127.0.0.1:8000`. You'll find the Django welcome web page.

Figure 1.15 – The Django default page

We've done it! Django is running on Docker.

As a last detail, if you are using the terminal, you will find that the container never stops. That's because the web server, as a good server, is constantly running and waiting for requests until we tell it otherwise. Press *Ctrl + C* if you want to close it. In PyCharm, you should click on the red **Stop** square.

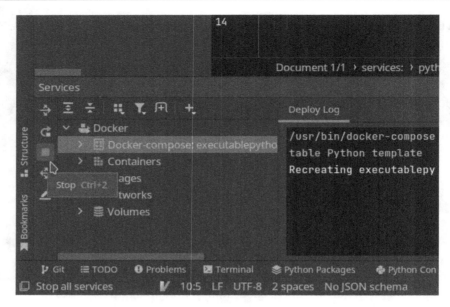

Figure 1.16 – Stopping Docker services via PyCharm and its integration

Summary

We have just acquired the skills to configure and build a Python project using Docker containers. We started with the basics, creating an image that runs a Python script and also installs all the dependencies we declared in `requirements.txt`. Then, we automated the creation of a Django project with a simple script and set up the development server.

On the other hand, to make container management easier, we have integrated an IDE into the flow, in our case, PyCharm. It gives us the possibility to launch some of the functionalities that we will use the most: building a custom image, executing a container composition (now we only have a service for Python), visualizing the log, and restarting and stopping containers. But let's not forget that all these tasks are accessible from the terminal, using `docker-compose`.

In the next chapter, we will build a complete project in Django with various databases, a web server, and other tools that we will need to build a complete project. In addition, we will integrate Django's configuration with Docker to facilitate its deployment with different configurations.

2

Creating a Django Project around Docker

In the previous chapter, we learned how to launch an application made in Python using a container system such as **Docker**. In addition, we created a project in **Django**, always with an eye on WebSockets for future real-time communication. At the moment, we only have a simple *executable*; we need to create a service architecture that complements Django. Important pieces such as a database to store and retrieve information, among other things (such as a fake mail server), will be useful for development. By configuring these tools, we will finish building an optimal working environment around Docker to then focus on the code.

We will also work on the communication and integration of environment variables to configure some aspects of the project through `docker-compose.yaml`. We will modify the critical elements of deployment, such as activating or deactivating the debug mode, changing the domain, indicating the path where the statics will be stored, and some other important particularities that differentiate a local development from a production server.

By the end of this chapter, we will have a fully integrated project with Django, ready for deployment on a test or real server, which will also be easy for other team members to pick up.

In this chapter, we'll be covering the following topics:

- Exploring the containers used for building our app
- Adding the Django service
- Configuring databases
- Connecting Django to a web server
- Adding a fake SMTP mail server
- Testing for correct operation

Technical requirements

The code for this chapter can be found at `https://github.com/PacktPublishing/Building-SPAs-with-Django-and-HTML-Over-the-Wire/tree/main/chapter-2`.

Exploring the containers used for building our app

Containers are processes isolated from your operating system. Docker allows us to modify them, add tools, execute scripts, and the like, all without leaving the memory space that Docker reserves for us while they are running. When we want to stop working with a container, we can stop it and any action we have performed will cease to exist. Of course, if we need to, we can save the changes in *volumes*. These are Docker virtual hard disks that can connect to any container that is mounted in folders; this is very useful in allowing the container to access project files or configurations.

We will use containers to create an environment that is easy to deploy, irrespective of the software or version of Python installed on the machine. In addition, we will be able to select the version for each software in a transparent way.

Let's start by extending `docker-compose.yaml` by adding the following services:

> **Django**: We will modify the Python service to a large extent. We will not only change its name but also add environment variables and a script that will perform management tasks.

- **PostgreSQL**: This will be the relational database that we will use. Although Django is database-agnostic, the framework itself recommends it (`https://bit.ly/3JUyfUB`) because PostgreSQL is rich in field types and has interesting extensions.

- **Caddy**: This is an excellent web server. It will be in charge of managing the domain, auto-renewing the SSL certificate, serving static documents, and being a reverse proxy to access the Django service.

- **Redis**: If you don't know it already, this is an in-memory database that works with a really fast key-value paradigm. We won't communicate directly with it but channels will when we open or close rooms. On the other hand, integrating it is a good idea, whether we use WebSockets, as it is an excellent caching system.

- **MailHog**: This is a simple SMTP server that will capture all the traffic sent by mail and display it on a graphical interface for users to visualize.

Using a Python service was enough to launch simple code, but now we must have a Docker service that integrates all the configurations and launches the Django server.

Adding the Django service

Within `docker-compose.yaml`, replace the entire `python` block with the following:

```
django:
    build:
        context: ./
        dockerfile: ./Dockerfile
    entrypoint: bash ./django-launcher.sh
    volumes:
        - .:/usr/src/app/
    environment:
        DEBUG: "True"
        ALLOWED_HOSTS: hello.localhost
        SECRET_KEY: mysecret
        DB_ENGINE: django.db.backends.postgresql
        DB_NAME: hello_db
        DB_USER: postgres
        DB_PASSWORD: postgres
        DB_HOST: postgresql
        DB_PORT: 5432
        DOMAIN: hello.localhost
        DOMAIN_URL: http://hello.localhost
        REDIS_HOST: redis
        REDIS_PORT: 6379
        DEFAULT_FROM_EMAIL: no-reply@hello.localhost
        STATIC_URL: /static/
        STATIC_ROOT: static
        MEDIA_URL: /media/
        EMAIL_HOST: mailhog
        EMAIL_USE_TLS: "False"
        EMAIL_USE_SSL: "False"
        EMAIL_PORT: 1025
        EMAIL_USER:
        EMAIL_PASSWORD:
    expose:
        - 8000
```

```
depends_on:
  - postgresql
  - redis
```

Let us go through each point:

- With `build`, as when we use `python`, we indicate that we generate the Python image that we have defined in `Dockerfile`:

```
build:
    context: ./
    dockerfile: ./Dockerfile
```

- As you can see, we have modified the command that will execute the service when it is up. In this case, it will be necessary to launch several commands, so we will choose to save it in a shell file that we will invoke:

```
entrypoint: bash ./django-launcher.sh
```

We'll create `django-launcher.sh` later, so we'll ignore it for now.

- We will mount and synchronize the volume, which is the service space, with the folder we are in. The structure of the current folder is as follows:

```
folder/project:folder/container
```

In the next snippet, the dot (`.`) represents the location of the project, the colon (`:`) is a separator, and `/usr/src/app/` is the path to the container where the project will be located:

```
volumes:
  - .:/usr/src/app/
```

- We define all the environment variables that we will later integrate with Django's configuration so that we can move between the local and a production server anytime:

```
environment:
  KEY: value
```

- We activate development mode:

```
DEBUG: True
```

- We indicate the domain that will be allowed (for the moment, we will use a fictitious one):

  ```
  ALLOWED_HOSTS: hello.localhost
  ```

- We define a cryptographic key:

  ```
  SECRET_KEY: mysecret
  ```

  When working locally, its complexity should not be important.

- We configure PostgreSQL as our database:

  ```
  DB_ENGINE: django.db.backends.postgresql
  ```

- We indicate a name for the database:

  ```
  DB_NAME: hello_db
  ```

  This will be created later with the PostgreSQL service that we have not yet added.

- We define a database user:

  ```
  DB_USER: postgres
  ```

- We add a password for the database:

  ```
  DB_PASSWORD: postgres
  ```

- We configure the database with the name of the service of the future database:

  ```
  DB_HOST: postgresql
  ```

- We indicate the PostgreSQL port (by default it uses port 5432):

  ```
  DB_PORT: 5432
  ```

- We add the domain we will use (do not add the protocol, such as https://):

  ```
  DOMAIN: hello.localhost
  ```

- We define the path to be used, which will match the *protocol* and the *domain*:

  ```
  DOMAIN_URL: http://hello.localhost
  ```

- We tell Django the address and the port of Redis, which is another service we have set up:

  ```
  REDIS_HOST: redis
  REDIS_PORT: 6379
  ```

- We give static a prefix:

```
STATIC_URL: /static/
```

- We create the folder where we will save the static files:

```
STATIC_ROOT: static
```

We will use the static folder in the same project

- We define the path for multimedia content:

```
MEDIA_URL: /media/
```

- We define all the configuration for the fake SMTP server:

```
DEFAULT_FROM_EMAIL: no-reply@hello.localhost
EMAIL_HOST: mailhog
EMAIL_USE_TLS: False
EMAIL_USE_SSL: False
EMAIL_PORT: 1025
EMAIL_USER:
EMAIL_PASSWORD:
```

We tell it to use the mailhog service, which does not exist yet, with port 1025. On a real server, it will probably be 25.

- The web server needs to access the Django server. We will open it on port 8000. There are two ways to do it, visible to all (ports) or only to the Docker subnet (expose). It only needs to be accessible to other services:

```
expose:
  - 8000
```

- Finally, please wait for the databases before launching. They do not yet exist, but they will be there soon:

```
depends_on:
  - postgresql
  - redis
```

Next, we will create a script to control the actions when we start the Django service.

Creating a Django launcher

It is good practice to keep track of the commands that will be executed each time the Django service is up. So, we'll create django-launcher.sh in the root of the project with the following content:

```sh
#!/bin/sh

# Collect static files
python3 manage.py collectstatic --noinput

# Apply database migrations
python3 manage.py migrate

# Start server with debug mode
python3 manage.py runserver 0.0.0.0 8000
```

In this way, every time we raise the Django service, we will get the static files, launch the new migrations, and raise the development server on port 8000.

We edit the hello_world/settings.py. We are going to import os to access the environment variables:

```python
import os
```

Next, we modify the following lines:

```python
SECRET_KEY = os.environ.get("SECRET_KEY")
DEBUG = os.environ.get("DEBUG", "True") == "True"
ALLOWED_HOSTS = os.environ.get("ALLOWED_HOSTS").split(",")
DATABASES = {
    "default": {
        "ENGINE": os.environ.get("DB_ENGINE"),
        "NAME": os.environ.get("DB_NAME"),
        "USER": os.environ.get("DB_USER"),
        "PASSWORD": os.environ.get("DB_PASSWORD"),
        "HOST": os.environ.get("DB_HOST"),
        "PORT": os.environ.get("DB_PORT"),
    }
}
```

```python
STATIC_ROOT = os.environ.get("STATIC_ROOT")
STATIC_URL = os.environ.get("STATIC_URL")
MEDIA_ROOT = os.path.join(BASE_DIR, "media")
MEDIA_URL = os.environ.get("MEDIA_URL")

DOMAIN = os.environ.get("DOMAIN")
DOMAIN_URL = os.environ.get("DOMAIN_URL")
CSRF_TRUSTED_ORIGINS = [DOMAIN_URL]

"""EMAIL CONFIG"""
DEFAULT_FROM_EMAIL = os.environ.get("EMAIL_ADDRESS")
EMAIL_USE_TLS = os.environ.get("EMAIL_USE_TLS") == "True"
EMAIL_HOST = os.environ.get("EMAIL_HOST")
EMAIL_PORT = os.environ.get("EMAIL_PORT")
EMAIL_HOST_USER = os.environ.get("EMAIL_HOST_USER")
EMAIL_HOST_PASSWORD = os.environ.get("EMAIL_HOST_PASSWORD")
STATIC_ROOT = os.environ.get("STATIC_ROOT")
STATIC_URL = os.environ.get("STATIC_URL")
MEDIA_ROOT = os.path.join(BASE_DIR, "media")
MEDIA_URL = os.environ.get("MEDIA_URL")

DOMAIN = os.environ.get("DOMAIN")
DOMAIN_URL = os.environ.get("DOMAIN_URL")
CSRF_TRUSTED_ORIGINS = [DOMAIN_URL]

"""EMAIL CONFIG"""
DEFAULT_FROM_EMAIL = os.environ.get("EMAIL_ADDRESS")
EMAIL_USE_TLS = os.environ.get("EMAIL_USE_TLS") == "True"
EMAIL_USE_SSL = os.environ.get("EMAIL_USE_SSL") == "True"
EMAIL_HOST = os.environ.get("EMAIL_HOST")
EMAIL_PORT = os.environ.get("EMAIL_PORT")
EMAIL_HOST_USER = os.environ.get("EMAIL_HOST_USER")
EMAIL_HOST_PASSWORD = os.environ.get("EMAIL_HOST_PASSWORD")

CHANNEL_LAYERS = {
    "default": {
```

```
        "BACKEND": "channels_redis.core.RedisChannelLayer",
        "CONFIG": {
            "hosts": [(os.environ.get("REDIS_HOST"),
 os.environ.get("REDIS_PORT"))],
        },
    },
 }
```

With that, you will have perfectly integrated Django. If you have problems, feel free to copy the example material online:

https://github.com/PacktPublishing/Building-SPAs-with-Django-and-HTML-Over-the-Wire/tree/main/chapter-2

An application without a database has very limited use. So, we're going to give Django two databases: PostgreSQL and Redis. You'll soon understand why.

Configuring the databases

We continue adding services in docker-compose.yaml. After the Django service, we add the following configuration:

```
postgresql:
    image: postgres
    environment:
      POSTGRES_USER: postgres
      POSTGRES_PASSWORD: postgres
      POSTGRES_DB: hello_db
    volumes:
      - ./postgres_data:/var/lib/postgresql/data/
    expose:
      - 5432
```

In image: postgres, we are using the official PostgreSQL image. It will be automatically downloaded from the official repositories. Next, we configure the environment variables to indicate the user credentials (POSTGRES_USER and POSTGRES_PASSWORD) and the name of the database (POSTGRES_DB). The variables must match those declared in the Django service; otherwise, it will fail to connect.

It is important to keep a copy of the database, otherwise everything is lost when rebooting. `postgres_data:/var/lib/postgresql/data/` indicates that all the PostgreSQL content in the container is saved in the `postgres_data` folder. Finally, we expose the port (`5432`) that Django will use to connect.

Then, we add Redis, the other key-value database:

```
redis:
    image: redis:alpine
    expose:
      - 6379
```

It's as simple as that. We use the official image with the `alpine` label to make it as light as possible and expose port `6379`.

We already have Django and the databases ready. The next step is to connect Django to a web server that exposes the project and manages the SSL certificates automatically.

Connecting Django to a web server

We must have a gateway service that manages the static content. We will use Caddy for its simplicity.

Caddy is configured with a flat file named `Caddyfile`. We must create it and add the following content:

```
http://hello.localhost {
    root * /usr/src/app/

    @notStatic {
      not path /static/* /media/*
    }

    reverse_proxy @notStatic django:8000

    file_server
}

http://webmail.localhost {
    reverse_proxy mailhog:8025
}
```

With the first line, `http://hello.localhost`, we indicate the domain that we will use. As we are in a development environment, we will indicate the *http* protocol instead of *https*. Next, with `root * /usr/src/app/` and `file_server`, we're telling Caddy to expose static files (images, CSS files, JavaScript files, and so on) because that's not Django's job. Finally, we reverse proxy the `django` service on port `8000`, ignoring its static or media routes to avoid conflicts.

The second block is again a reverse proxy that will point to the fake SMTP mail interface with the `webmail.localhost` domain.

After leaving the configuration ready, we have to create the service. We add `docker-compose.yaml` into the Caddy service:

```
caddy:
    image: caddy:alpine
    ports:
        - 80:80
        - 443:443
    volumes:
        - ./Caddyfile:/etc/caddy/Caddyfile
        - ./caddy_data:/data
    depends_on:
        - django
```

As with Redis, we use the official image in its alpine version: `image: caddy:alpine`. We open ports `80` and `443` publicly so that any visitor can access the site. The next thing is to connect two volumes: the `Caddyfile` configuration file with the internal one of the container (`./Caddyfile:/etc/caddy/Caddyfile`) and the Caddy information with a folder that we will save in the project named `caddy_data` (`./caddy_data:/data`).

The next step will be to add a mail server to test that future mail is received correctly by users. In addition, we will test that the rest of the services work as they should.

Adding a fake SMTP mail server

At the end of the `docker-compose.yaml`, we add the last service:

```
mailhog:
    image: mailhog/mailhog:latest
    expose:
        - 1025
        - 8025
```

The ports used will be `1025` for Django to connect to the SMTP server and `8025` for the web interface via the `webmail.localhost` domain because Caddy will act as a reverse proxy.

Now that we have added all the containers, it's time to test whether the containers run and work with each other.

Testing for correct operation

Finally, we pull up all the services from `docker-compose.yaml` to test whether the containers run and work with each other.

Caddy and Django

Caddy and Django are easy to check, as when you enter the `hello.localhost` domain, you will see Django fully functioning with its welcome page:

django   View <u>release notes</u> for Django 4.0

The install worked successfully!
Congratulations!
You are seeing this page because <u>DEBUG=True</u> is in your settings file and you have not configured any URLs.

Figure 2.1 – Django running under the hello.localhost domain

We know that Django has connected to PostgreSQL because we can see in the log how it has applied the migrations:

```
Running migrations:
Applying contenttypes.0001_initial... OK
Applying auth.0001_initial... OK
Applying admin.0001_initial... OK
Applying admin.0002_logentry_remove_auto_add... OK
Applying admin.0003_logentry_add_action_flag_choices... OK
Applying contenttypes.0002_remove_content_type_name... OK
Applying auth.0002_alter_permission_name_max_length... OK
...
```

MailHog

MailHog is simple because when you enter the webmail.localhost domain, you will see the web interface with an empty inbox:

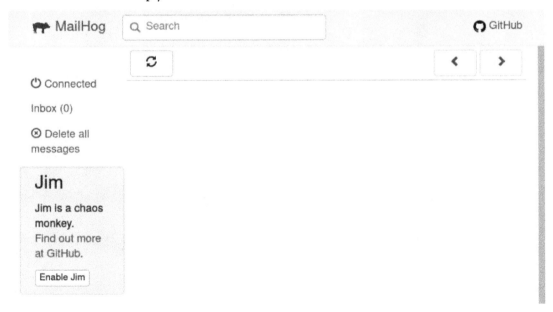

Figure 2.2 – MailHog WebMail with an empty inbox

And finally, with Redis, we only have to make sure that there are no errors in the log. Otherwise, it will be silent.

Summary

In the previous chapter, we were able to run Python in complete isolation from the operating system, including its dependencies. It wasn't much different from creating a virtual environment. But in this iteration, we've taken it up a notch by incorporating all the external software into Django. Containers have become the backbone of the site, incorporating elements as important as databases and the web server itself. What's more, integration with Django is not merely decorative, as the most critical configurations originate in the Docker environment variables that directly affect `settings.py`. Right now, if we wanted to, we could deploy the site on any server that has Docker installed with just one command. We can find harmony and architecture in every line of Docker.

We are now ready to dive into asynchrony, WebSockets, channels, and real-time requests.

Part 2:
WebSockets
in Django

In this part, we will work with a WebSockets server using Django Channels. We will learn how to send messages in real-time from the backend, from plain text, JSON, to complex HTML structures, and how to send messages from the frontend. We will also learn how to use the database to monitor connections, discriminate messages to groups of users or broadcast messages.

In this part, we cover the following chapters:

3

Adding WebSockets to Django

Channels is a Django extension that allows us to use protocols other than HTTP. The Django team, knowing the existing limitations of including other protocols, had to create a new server called Daphne that is natively compatible with the **Asynchronous Server Gateway Interface (ASGI)**, an update of the **Web Server Gateway Interface (WSGI)**. Without Channels, it would be impossible to have the WebSockets protocol.

You may wonder why the migration from WSGI to ASGI is so important. First, we need to understand what communication interfaces are. When we want to serve a Python site, be it Django or any other framework, we need to run software capable of keeping an instance active and mediating with any web server that makes requests. There are many interfaces for a web server to understand Python, but the most recommended is the WSGI specification, a Python standard for communication between a web server (Apache, Nginx, Caddy, etc.) and any Python web app or frameworks/applications (Django, Flask, FastAPI, etc.). Unfortunately, there is a limitation.

When we want to use HTTP with another protocol (WebSockets, etc.) it is not possible; HTTP was not designed to work with other protocols. The alternative is to use the ASGI specification, which will allow us to accept different protocols and asynchronous requests, split tasks into events, and keep a request alive after its response. In addition, it is compatible with applications using WSGI, so it will work perfectly with existing applications. That's why Daphne (remember that it's Django's new ASGI server) is a spiritual successor as it maintains compatibility and extends Django possibilities. And this is the reason why we included the dependencies of **daphne**, **asgiref**, and **channels** in *Chapter 2, Creating a Django Project around Docker*, when we created the Docker container; they are the minimum tools required to be able to work with the WebSockets protocol.

In this chapter, we'll learn how to activate Channels, configure Django to use Daphne, create a communication channel between the backend and frontend, send different formats (plain text, JSON, and HTML), where Django will generate the HTML blocks that will be displayed in the frontend. We will also delve into other essential Channels components such as **scope**, **consumers**, **routing**, and events.

By the end of this chapter, we will acquire the minimum skills required to establish bidirectional communication between the server and the client using Django.

We will be covering the following topics in this chapter:

- Creating our first page with Daphne

- Listening to events with consumers

- Sending plain text from the backend

- Sending and receiving messages in JSON

- Rendering HTML in the backend

Technical requirements

The code for this chapter can be found at `https://github.com/PacktPublishing/Building-SPAs-with-Django-and-HTML-Over-the-Wire/tree/main/chapter-3`.

Creating our first page with Daphne

Let's begin by activating all the real-time capabilities that Channels has to offer. So, we are going to enable Channels in the configuration, change the WSGI server for ASGI, and build a minimal HTML page to check that everything is running correctly. We will continue the application we started in *Chapter 2, Creating a Django Project around Docker*. If you don't have the code, you can use the example available in the following repository: `https://github.com/PacktPublishing/Building-SPAs-with-Django-and-HTML-Over-the-Wire/tree/main/chapter-2`.

Let's begin:

1. The first thing we are going to do is to activate Channels. To do this, we open `hello_world/settings.py`, and under `INSTALLED_APPS`, we add `app.simple_app` (put it at the end) and `channels` (put it first):

    ```
    INSTALLED_APPS = [
        ' channels', # New line
        'django.contrib.admin',
        'django.contrib.auth',
        'django.contrib.contenttypes',
        'django.contrib.sessions',
        'django.contrib.messages',
        'django.contrib.staticfiles',
        'app.simple_app', #New line
    ]
    ```

2. As we learned in the introduction, Channels needs an ASGI-compatible server to work. If we don't meet this requirement, Django won't even be able to get up. That's why we have to indicate where the file with the configuration for Daphne, or any other server that manages the interface, is located. In our case, we will go to the end of settings.py and add the following line:

ASGI_APPLICATION = "hello_world.asgi.application"

3. As the project is inside the app folder, Django may not be able to handle imports properly. To fix this, we'll change the name of our application. We open app/simple_app/apps.py and leave it as follows:

```
from django.apps import AppConfig


class SimpleAppConfig(AppConfig):
    default_auto_field = 'django.db.models.
        BigAutoField'.
    name = 'app.simple_app' # Update
```

4. We will next create a basic HTML page, where JavaScript will interact with the future WebSockets server.

Inside app/simple_app/, we create the templates folder, and inside it we create a new file called index.html. The full path would be app/simple_app/templates/index.html. We include the following content:

```
<html lang="en">
<head>
    <meta charset="UTF-8">
    <meta name="viewport" content="width=device-width,
      user-scalable=no, initial-scale=1.0, maximum-
        scale=1.0, minimum-scale=1.0">
    <title> Index </title>
</head>
<body>
    <h1> Hello Django! <h1>
</body>
</html>
```

5. Next, let's create a view that serves the HTML file we just created. We open `app/simple_app/views.py` and create the view `index`. In `return`, we'll tell it to respond with the content of the template:

```python
from django.shortcuts import render


def index(request):
    return render(request, 'index.html', {})
```

6. Now we just need to give a route. We go into `hello_world/urls.py`. Import the view and add a new `path`:

```python
from django.contrib import admin
from django.urls import path
from app.simple_app import views


urlpatterns = [
    path('', views.index, name='index'), # New line
    path('admin/', admin.site.urls),
]
```

7. We pull up Docker and go to `http://hello.localhost/`. It will respond with a minimalist message:

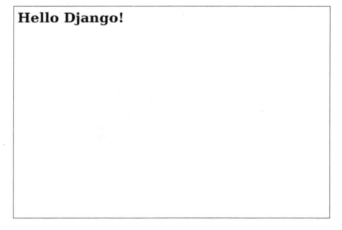

Hello Django!

Figure 3.1 – Displaying a static HTML page on the domain hello.localhost

Congratulations! You have built a web page with Daphne and Django that is ready to handle WebSockets connections.

The next step is to build and learn about the concept of Consumers, Channels' way of handling events. We'll connect the frontend with a simple Consumer that will act as an intermediary to communicate between Django and the JavaScript WebSocket client.

Listening to events with consumers

We are going to build an example where we can send messages between a backend and frontend in real time. We will need an intermediary to listen to both and make decisions. Channels comes ready with a specially prepared tool called a consumer. It is a series of functions that are invoked when an event is triggered by a WebSocket client.

Here you can see a minimal consumer structure for WebSockets:

```python
from channels.generic.websocket import WebsocketConsumer

class NameConsumer(WebsocketConsumer):

    def connect(self):
        """Event when client connects"""
        # Informs client of successful connection
        self.accept()

    def disconnect(self, close_code):
        """Event when client disconnects"""
        pass

    def receive(self, text_data):
        """Event when data is received"""
        pass
```

Let us go through what each of these events in the preceding code snippet does:

- connect: Asks for permission to connect. If we accept, we will assign a group. More on this later.

- receive: Sends us information and we decide what to do with it. It may trigger some actions, or we may ignore it; we are not obliged to process it.

- disconnect: Notifies us that it is going to close the connection. It is important to be aware in advance if the connection is going to close.

Being technical, Consumers are an abstraction for making event-driven applications using ASGI. They allow us to execute the code we need when a change occurs. If you have worked with JavaScript events, you will be very familiar with them. And if you haven't, you could think of them as Channels views, except, instead of being triggered by requests to the URLs we link to the views, they are actions that execute functions.

Django is able to capture client events. Channels knows whether a new client has connected, disconnected, or sent information. However, the backend has a very powerful ability: to select the recipients of the information. When a client asks to connect, Django must assign the client a group or channel. The client can be left alone or grouped with other clients, depending on the need. When the time comes to send a message, we must decide who will receive it—a specific client or a group.

In what situation will I need to send data to a client or a group? Let's use the example of a chat. When I write a private message, Django will send the new text to a specific client, that is the specific user I'm talking to. On the other hand, if I am in a group with other users, when I write a message, the backend will send the text only to a specific selection, that is all the users subscribed to the chat room. The frontend does not decide whether it receives new information or who the recipients are; it is the consumer who is the master of who moves the available information.

Now that we have seen how consumers can be useful, let's look at the role they play when sending plain text to a server/client. We will explore how a client can capture information received from a channel.

Sending plain text from the backend

Let's build the first minimalistic consumer that greets a WebSocket client when it connects. Later, we will add complexity and other actions.

All the code for the example can be found at `https://github.com/PacktPublishing/Building-SPAs-with-Django-and-HTML-Over-the-Wire/tree/main/chapter-3/Sending%20plain%20text`:

1. We create `app/simple_app/consumers.py` with the following content:

    ```
    # app/simple_app/consumers.py
    from channels.generic.websocket import WebsocketConsumer

    class EchoConsumer(WebsocketConsumer):

        def connect(self):
            """Event when client connects"""

            # Informs client of successful connection
            self.accept()
    ```

```
        # Send message to client
        self.send(text_data="You are connected by
WebSockets!")

    def disconnect(self, close_code):
        """Event when client disconnects"""
        pass

    def receive(self, text_data):
        """Event when data is received"""
        pass
```

Let's explain each element of the file we have just created:

- With `from channels.generic.websocket import WebsocketConsumer`, we import the consumer object for WebSockets.

- We declare the Consumer with `class EchoConsumer(WebsocketConsumer):` and call it EchoConsumer.

- You need at least three functions that are the events that will be triggered by the frontend actions: `connect`, `disconnect`, and `receive`. We are going to focus on `connect`.

- When clients connect, the first thing we'll do is confirm the connection to establish future communication between Django and the client. We use `self.accept()`.

- Finally, we will send a message with `self.send`, the string `"You are connected by WebSockets!"`. That way every client that connects will receive a greeting.

2. Now we need to assign the consumer a route for a WebSockets client to connect to. For this, we add the routes on the ASGI server that connects to the consumers. We open `hello_world/asgi.py` and update with the path `/ws/echo/` pointing to EchoConsumer:

```
# hello_world/asgi.py
import os
from django.core.asgi import get_asgi_application
from channels.auth import AuthMiddlewareStack
from channels.routing import ProtocolTypeRouter,
    URLRouter
from django.urls import re_path
from app.simple_app.consumers import EchoConsumer
```

```
    os.environ.setdefault('DJANGO_SETTINGS_MODULE',
        'hello_world.settings')


application = ProtocolTypeRouter({
    # Django's ASGI application to handle traditional
    HTTP requests
    "http": get_asgi_application(),
    # WebSocket handler
    "websocket": AuthMiddlewareStack(
        URLRouter([
            re_path(r"^ws/echo/$", EchoConsumer.
                as_asgi()),
        ])
    ),
})
```

In `asgi.py`, you will find all the configurations that you can apply to the ASGI server:

- Since ASGI loads before Django, it can't know the path to Django's own configuration file. We must preempt this with `os.environ.setdefault('DJANGO_SETTINGS_MODULE', 'hello_world.settings')`.

- In `application`, we configure all routes, whether HTTP or any other protocol. We use the `ProtocolTypeRouter` object to indicate the type and destination of each route.

- It will still be necessary to use traditional HTTP requests: loading pages via HTTP, managing sessions, cookies, and other particularities. For this task, we include `"http": get_asgi_application()` inside `ProtocolTypeRouter`.

- Finally, we include the consumer path with `re_path(r'ws/echo/$', consumers.EchoConsumer.as_asgi())`. Now the path to connect to EchoConsumer is `/ws/echo/`.

3. Next, we connect a WebSockets client to Django.

   We go to `app/simple_app/templates/index.html`, where we are going to define with JavaScript a WebSockets client that connects to the path we just created:

```
{# app/simple_app/templates/index.html #}
<! doctype html>
<html lang="en">
<head>
    <meta charset="UTF-8">
```

```
    <meta name="viewport" content="width=device-width,
        user-scalable=no, initial-scale=1.0, maximum-
            scale=1.0, minimum-scale=1.0">
    < title> Index </title>
</head>
<body>
    <h1>Hello Django!</h1>

    <!-- Place where we will display the connection
    message. -->
    <h2 id="welcome"></h2>

    <script>
        // Connect to WebSockets server (EchoConsumer)
        const myWebSocket = new WebSocket("ws://{{
            request.get_host }}/ws/echo/");

        // Event when a new message is received by
        WebSockets
        myWebSocket.addEventListener("message",
            (event) => {
            // Display the message in '#welcome'.
            document.querySelector("#welcome").
                textContent = event.data;
        });
    </script>
</body>
</html>
```

We've created a minimalistic WebSockets client in JavaScript to listen to everything we receive from the backend:

- We define an HTML tag to display the welcome message that the backend will send us. Adding `<h2 id="welcome"></h2>` will be enough; later, we will fill it with JavaScript.

- We connect with new `WebSocket()` to `/ws/echo/`. The address must contain the following structure: `protocol://domain/path`. In our case, it will be `ws://hello.localhost/ws/echo/`.

- The `message` event will be fired when the backend sends a message. As soon as we connect to Django, we'll receive the message we've programmed and then display it in `<h2>`.

WebSocket secure protocol

We can use the `ws://` protocol, where information is sent in plain text, or `wss://` to maintain a secure connection. The difference is similar to using `http://` or `https://`. We will change the protocol to secure when it is in production or when we can provide an SSL certificate; while we are developing, it is not necessary.

Open in your favorite browser the address `http://hello.localhost`.

Hello Django!
You are connected by WebSockets!

Figure 3.2 – Sending plain text from the backend a plain text ("You are connected by WebSockets") and rendering the message in an HTML element, below the title

We just learned how to send asynchronous plain text from the backend to the frontend via WebSockets. OK! It's not very spectacular; it's sent at the very moment we connect. However, as we have it built, we can send messages whenever we need to.

Let's make it more interesting: do we synchronize the time for all our visitors? In other words, send the same information to all connected clients in real-time. Of course!

We go back to `app/simple_app/consumers.py`. We will create an infinite loop wherein every second a text is sent to the frontend, specifically, the current time. We'll use **threading** to create a background task and not produce any interruptions:

```python
# app/simple_app/consumers.py
from channels.generic.websocket import WebsocketConsumer
from datetime import datetime # New line
import time # New line
import threading # New line


class EchoConsumer(WebsocketConsumer):
```

```python
def connect(self):
    """Event when client connects"""

    # Informs client of successful connection
    self.accept()

    # Send message to client
    self.send(text_data="You are connected by
        WebSockets!")

    # Send message to client every second
    def send_time(self): # New line
        while True:
            # Send message to client
            self.send(text_data=str(datetime.now().
                Strftime("%H:%M:%S")))
            # Sleep for 1 second
            time.sleep(1)
    threading.Thread(target=send_time, args=(self,)).
        start() # New line

def disconnect(self, close_code):
    """Event when client disconnects"""
    pass

def receive(self, text_data):
    """Event when data is received"""
    pass
```

Now, open the same address hello.localhost in different tabs or browsers; you'll see how they show the exact same time. All clients are synchronized, receiving the same. No waiting, no need to ask the backend.

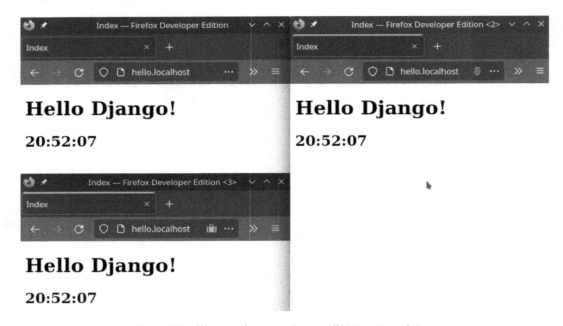

Figure 3.3 – Showing the same time to all visitors in real time

The power of real time before your eyes. Maybe it's inspiring your imagination; the possibilities are endless: an election system? An auction site? Notifications? Taxi locator? Food order? Chat? We'll keep exploring with small projects.

In the following sections, we will continue to learn how to send messages from the backend in different formats, such as JSON or HTML.

Sending JSON from the backend

We are going to send content in JSON format from the backend and consume it at the frontend. In addition, we will give the code a reusable structure that will be useful throughout the book.

All the code for the example can be found at `https://github.com/PacktPublishing/Building-SPAs-with-Django-and-HTML-Over-the-Wire/tree/main/chapter-3/Sending%20JSON`.

We have a consumer type adapted for the purpose of sending or receiving JSON called `JsonWebsocketConsumer`. It is the same as `WebsocketConsumer` except for two differences:

- We need to add the `send_json` function to encode to JSON:

```
book = {
    'title': 'Don Quixote',
```

```
       author': 'Miguel de Cervantes'.
   }
   self.send_json(content=book)
```

- We have a new event, called `receive_json`, which automatically decodes JSON when a message is received from the client:

```
def receive_json(self, data):
    """Event when data is received"""
    pass
```

To exemplify how we can send content in JSON format, we will create a Bingo project.

Example project – Generating a ticket of random numbers

When a client connects, a ticket with a series of random numbers will be generated and delivered to them via WebSockets. Then, every so often, we will send a random ball from Django. When the player has all the numbers, we will display a winner message.

Let's get started:

1. We add to `app/simple_app/consumers.py` a new consumer extending from `JsonWebsocketConsumer`:

```
from channels.generic.websocket import
JsonWebsocketConsumer

class BingoConsumer(JsonWebsocketConsumer):

    def connect(self):
        self.accept()

    def disconnect(self, close_code):
        """Event when client disconnects"""
        pass

    def receive_json(self, data):
        """Event when data is received"""
        Pass
```

2. We generate a ticket of five numbers between 1 and 10. We will avoid repetitions with set ():

```
class BingoConsumer(JsonWebsocketConsumer):

    def connect(self):
        self.accept()
        ## Send numbers to client
        # Generates numbers 5 random numbers,
            approximately, between 1 and 10
        random_numbers = list(set([randint(1, 10) for
            _ in range(5)])))
        message = {
            ' action': 'New ticket',
            ' ticket': random_numbers
        }
        self.send_json(content=message)

    def disconnect(self, close_code):
        """Event when client disconnects"""
        pass

    def receive_json(self, data):
        """Event when data is received"""
        pass
```

To inform the client what kind of action we are going to perform, we include in the JSON to send action. Separately, in ticket, we include the list of numbers.

3. Edit hello_world/asgi.py and add the path /ws/bingo/ pointing to BingoConsumer. Don't forget to import it. Now we have a new endpoint to feed the future WebSockets client. It's time to create the HTML:

```
# hello_world/asgi.py
import os
from django.core.asgi import get_asgi_application
from channels.auth import AuthMiddlewareStack
from channels.routing import ProtocolTypeRouter,
URLRouter
from django.urls import re_path
```

```
from app.simple_app.consumers import EchoConsumer,
BingoConsumer # Update

os.environ.setdefault('DJANGO_SETTINGS_MODULE',
    'hello_world.settings')

application = ProtocolTypeRouter({
    # Django's ASGI application to handle traditional
    HTTP requests
    "http": get_asgi_application(),
    # WebSocket handler
    "websocket": AuthMiddlewareStack(
        URLRouter([
            re_path(r"^ws/echo/$", EchoConsumer.
                as_asgi()),
            re_path(r"^ws/bingo/$", BingoConsumer.
                as_asgi()), # New line
        ])
    ),
})
```

The consumer is prepared to send a ticket with random numbers to every client that connects. The next step will be to prepare the frontend to receive it and render it in the appropriate HTML element.

Receiving JSON in the frontend

The goal will be to receive an asynchronous JSON from the backend that JavaScript will detect in an event. With the data, we will show in the HTML the information to the visitor:

1. We create a new HTML in app/simple_app/templates/bingo.html that will contain all the frontend:

```
{# app/simple_app/templates/bingo.html #}
<! doctype html>
<html lang="en">
<head>
    <meta charset="UTF-8">
    <meta name="viewport"
```

```
                content="width=device-width, user
                    scalable=no, initial-scale=1.0, maximum-
                        scale=1.0, minimum-scale=1.0">
        <title>Bingo</title>
</head>
<body>
    <h1>Bingo</h1>
    <h2>Ball: <span id="ball"></span></h2>.
    <h2 id="ticket"></h2>

    <script>
        // Connect to WebSockets server
        (BingoConsumer)
        const myWebSocket = new WebSocket("ws://{{
            request.get_host }}/ws/bingo/");
        let ticket = [];

        // Event when a new message is received by
            WebSockets
        myWebSocket.addEventListener("message",
            (event) => {
            const myData = JSON.parse(event.data);
            switch (myData.action) {
                case "New ticket":
                    // Save the new ticket
                    ticket = myData.ticket;
                    // Render ticket
                    document.getElementById("ticket").
                        textContent = "Ticket: " +
                            ticket.join(", ");
                    break;
            } }
        });
    </script>
</body>
</html>
```

2. We will need a view for the template we have created. We add to `app/simple_app/views.py` the following function:

```
from django.shortcuts import render

def index(request):
    return render(request, 'index.html', {})

def bingo(request): # New function
    return render(request, 'bingo.html', {})
```

3. In `hello_world/urls.py`, we include the `/bingo/` path:

```
from django.contrib import admin
from django.urls import path
from app.simple_app import views

urlpatterns = [
    path('', views.index, name='index'),
    path('bingo/', views.bingo, name='bingo'), # New line
    path('admin/', admin.site.urls),
]
```

And with this change, ticket generation will be ready.

4. When we enter `http://hello.localhost/bingo/`, we will see a ticket of random numbers that will only be given to us:

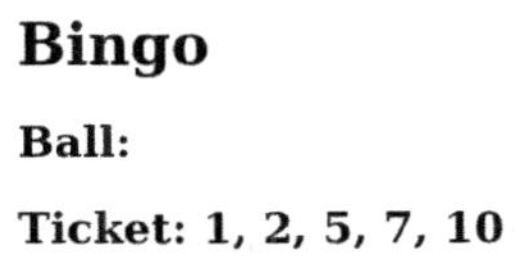

Bingo

Ball:

Ticket: 1, 2, 5, 7, 10

Figure 3.4 – Backend returning a set of random numbers when we connect via WebSockets

Currently, the consumer returns a JSON like the following to any client that connects to `/ws/bingo/`:

```
{
" action": " New ticket "
" ticket": [1, 2, 3...] // Random numbers
}
```

JavaScript waits, listening. If it receives a JSON whose "action" is "New ticket", it will store the entire contents of "ticket" in the ticket variable. Finally, the JSON displays the HTML:

```javascript
myWebSocket.addEventListener("message", (event) => {
const myData = JSON.parse(event.data);
switch (myData.action) {
case "New ticket":
// Save the new ticket
ticket = myData.ticket;
// Render ticket
document.getElementById("ticket"). textContent =
    "Ticket: " + ticket.join(", ");
                        break;
}
}
});
```

We now have an automated system for each customer to generate their own set of numbers to play. The next step will be to send all customers the same random number to represent the ball.

Example project – Checking for matching numbers

The next milestone is to send random numbers, on a recurring basis, from the backend for the frontend to check for matches. It's time to mix the balls!

1. We create a thread that generates random numbers and sends them every second. We will call the action 'New ball':

```python
# app/simple_app/consumers.py
from channels.generic.websocket import WebsocketConsumer
from datetime import datetime
import time
import threading
from random import randint
from channels.generic.websocket import
JsonWebsocketConsumer


class EchoConsumer(WebsocketConsumer):
 # Echo Code
```

```python
class BingoConsumer(JsonWebsocketConsumer):

    def connect(self):
        self.accept()
        ## Send numbers to client
        # Generates numbers 5 random numbers,
approximately, between 1 and 10
        random_numbers = list(set([randint(1, 10) for
            _ in range(5)]))
        message = {
            ' action': 'New ticket',
            ' ticket': random_numbers
        }
        self.send_json(content=message)

        ## Send balls
        def send_ball(self):
            while True:
                # Send message to client
                random_ball = randint(1, 10)
                message = {
                    ' action': 'New ball',
                    ' ball': random_ball
                }
                self.send_json(content=message)
                # Sleep for 1 second
                time.sleep(1)

        threading.Thread(target=send_ball,
            args=(self,)). start()

    def disconnect(self, close_code):
        """Event when client disconnects"""
        pass
```

```python
def receive_json(self, data):
    """Event when data is received"""
    Pass
```

2. In the JavaScript event that listens to Django, we'll add the case to detect whether an action with "New ball" arrives:

```html
{# app/simple_app/templates/bingo.html #}
<! doctype html>
<html lang="en">
<head>
    <meta charset="UTF-8">
    <meta name="viewport"
          content="width=device-width, user-
            scalable=no, initial-scale=1.0, maximum-
            scale=1.0, minimum-scale=1.0">
    <title>Bingo</title>
</head>
< body>
    <h1>Bingo</h1>
    <h2>Ball: <span id="ball"></span></h2>.
    <h2 id="ticket"></h2>

    <script>
        // Connect to WebSockets server (BingoConsumer)
        const myWebSocket = new WebSocket("ws://{{
            request.get_host }}/ws/bingo/");
        let ticket = [];

        // Event when a new message is received by
        WebSockets
        myWebSocket.addEventListener("message",
            (event) => {
            const myData = JSON.parse(event.data);
            switch (myData.action) {
                case "New ticket":
                    // Save the new ticket
```

```
            ticket = myData.ticket;
            // Render ticket
            document.getElementById("ticket").
                textContent = "Ticket: " +
                    ticket.join(", ");
            break;
        case "New ball":
            // Get the ball number
            ball = myData.ball;
            // Check if ball is in the ticket
            and remove it
            ticket = ticket.map(item => item
                === ball ? "X" : item);
            // Render ticket
            document.getElementById("ticket").
                textContent = "Ticket: " +
                    ticket.join(", ");
            // Render ball
            document.getElementById("ball").
                textContent = ball;
            // Check if we have a winner
            if (ticket.find(number => number
                !== "X") === undefined) {
                // We have a winner
                document.getElementById
                    ("ticket"). textContent =
                        "Winner!";
            }
            break;
        }
    });

</script>
</body>
</html>
```

3.   If so, we search the array of tickets for a match and replace the number with an X:

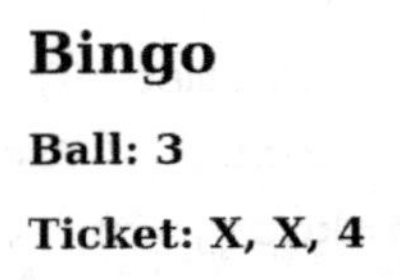

Bingo

Ball: 3

Ticket: X, X, 4

Figure 3.5 – If any of the balls matches a number on the ticket, we replace it with an X

And how do we know we have won? If all the contents of my ticket are X: game over!

Bingo

Ball: 1

Winner!

Figure 3.6 – All ticket numbers have been crossed out, so we display a winner message

So far, all that the frontend has done is listen and obey like a good child. However, they're now mature enough to be heard. From JavaScript, we're going to communicate with Django by making requests or sending information, and the backend will respond in two ways: either with JSON (as we've learned in this section) or with rendered HTML.

Rendering HTML in the backend

In this chapter, we will take the first steps on the principles of HTML over WebSockets. The backend will be in charge of rendering the HTML, taking the responsibility away from JavaScript and simplifying its tasks. On the other hand, we will avoid the need to incorporate a framework such as React, Vue, or Angular, and an API for an HTTP client.

The goal will be to build a **body mass index (BMI)** calculator for adults using the metric system. All calculations and HTML creation will be a Django task.

All the code for the example can be found at `https://github.com/PacktPublishing/Building-SPAs-with-Django-and-HTML-Over-the-Wire/tree/main/chapter-3/Rendering%20HTML`.

We will ask for the height, in centimeters, and the weight, in kilograms. The formula to get it is *weight (kg) / (height (m))²*.

In Python, it would be translated as follows: `weight / (height ** 2)`.

And its result will indicate the status:

Below 18.5	Underweight
18.5 - 24.9	Normal
25.0 - 29.9	Overweight
30.0 and Above	Obesity

Table 1.1 – BMI status

For example, if I weigh 78 kg and I am 180 cm tall (or 1.8 m), the calculation would be 78 / (1.8 **
2), resulting in 24. I would be in a *Normal* state, just one point away from *Overweight* – I think this
is life's warning for me to give up my daily chocolate desserts:

1. I start by adding a consumer named BMIConsumer in app/simple_app/
    consumers.py:

    ```
    from django.template.loader import render_to_string


    class BMIConsumer(JsonWebsocketConsumer):

        def connect(self):
            self.accept()


        def disconnect(self, close_code):
            """Event when client disconnects"""
            pass


        def receive_json(self, data):
            """Event when data is received"""
            height = data['height'] / 100
            weight = data['weight']
            bmi = round(weight / (height ** 2), 1)
            self.send_json(
                content={
                        "action": "BMI result",
                        "html": render_to_string(
                            "components/_bmi_result.html",
                            {"height": height, "weight":
    ```

```
                          weight, "bmi": bmi}
            )
        }
    )
```

For the first time, we will receive information from the client. The client will provide us with the values of a future form with weight and height, and in return we will return the HTML ready to display.

Let's explain what's going on in the preceding code snippet:

- We import the Django function to render HTML: `from django.template.loader import render_to_string`.

- Everything happens inside the `receive_json` function. With `data['height']` and `data['weight']`, we collect two pieces of data that we will send from JavaScript.

- Calculate the index with `round(weight / (height ** 2), 1)`.

- We send back to the client a JSON with two fields: `"action"`, where we simply inform the client what action to take, and `"html"` with the HTML generated from **render_to_string**.

2. Edit `hello_world/asgi.py` and add the `/ws/bmi/` path pointing to `BMIConsumer`.

```python
# hello_world/asgi.py
Import
from django.core.asgi import get_asgi_application
from channels.auth import AuthMiddlewareStack
from channels.routing import ProtocolTypeRouter,
URLRouter
from django.urls import re_path
from app.simple_app.consumers import EchoConsumer,
BingoConsumer, BMIConsumer # Update

os.environ.setdefault('DJANGO_SETTINGS_MODULE', 'hello_
world.settings')

application = ProtocolTypeRouter({
    # Django's ASGI application to handle traditional
    HTTP requests
    "http": get_asgi_application(),
    # WebSocket handler
    "websocket": AuthMiddlewareStack(
```

```
URLRouter([
    re_path(r"^ws/echo/$", EchoConsumer.
        as_asgi()),
    re_path(r"^ws/bingo/$", BingoConsumer.
        as_asgi()),
    re_path(r"^ws/bmi/$", BMIConsumer.
        as_asgi()), # New line
])
),
})
```

3. We create a new HTML in `app/simple_app/templates/bmi.html` that will contain the form and the JavaScript that will send the information:

```
{# app/simple_app/templates/bmi.html #}
<! doctype html>
<html lang="en">
<head>
    <meta charset="UTF-8">
    <meta name="viewport"
        content="width=device-width, user-
            scalable=no, initial-scale=1.0, maximum-
                scale=1.0, minimum-scale=1.0">
    <title>BMI Calculator</title>.
</head>
<body>
    <h1>BMI Calculator</h1>
    <label for="height"> Height (cm):
        <input type="text" name="height" id="height">
    </label>
    <label for="weight"> Weight (kg)
        <input type="text" name="weight" id="weight">
    </label>
    <input type="button" id="calculate" value=
        "Calculate">
    <div id="result"></div>
```

```html
<script>
    // Connect to WebSockets server
    (BingoConsumer)
    const myWebSocket = new WebSocket("ws://{{
        request.get_host }}/ws/bmi/");

    // Event when a new message is received by
    WebSockets
    myWebSocket.addEventListener("message",
        (event) => {
        const myData = JSON.parse(event.data);
        switch (myData.action) {
            case "BMI result":
                document.getElementById("result").
                    innerHTML = myData.html;
                break;
        }
    });

    document.querySelector('#calculate').
        addEventListener('click', () => {
        const height = parseFloat(document.
            querySelector('#height'). value);
        const weight = parseFloat(document.
            querySelector('#weight'). value);
        myWebSocket.send(JSON.stringify({
            height: height,
            weight: weight
        }));
    });

</script>
</body>
</html>
```

The mechanism is simple. When the **Calculate** button is clicked, it will get the input data and send it to the backend. If it receives an input with the `"action"` `"BMI results"`, it will inject the HTML in the appropriate place.

4. We will need a view for the template we have created. We add to `app/simple_app/views.py` the `bmi` function that points to the template:

    ```python
    from django.shortcuts import render


    def index(request):
        return render(request, 'index.html', {})


    def bingo(request):
        return render(request, 'bingo.html', {})


    def bmi(request): # New function
        return render(request, 'bmi. html', {})
    ```

5. In `hello_world/urls.py`, we include the `/bmi/` path:

    ```python
    from django.contrib import admin
    from django.urls import path
    from app.simple_app import views


    urlpatterns = [
        path('', views.index, name='index'),
        path('bingo/', views.bingo, name='bingo'),
        path('bmi/', views.bmi, name='bmi'), # New line
        path('admin/', admin.site.urls),
    ]
    ```

Now when we enter `http://hello.localhost/bmi/`, we will be shown the website with the form:

BMI Calculator

Height (cm): [_____]   Weight (kg) [_____]
[Calculate]

Figure 3.7 – The form is displayed ready for use

6. We only need the HTML component that we will use to display the content. We create in `app/simple_app/templates/components/_bmi_result.html` a document with the following content:

```
<p><strong> Weight</strong> {{ weight }} Kg</p>
<p><strong>Height</strong> {{ height }} m</p>
<p><p><strong>BMI</strong> {{ bmi }}< /p>
{% if bmi < 18.5 %}
<p>Underweight</p>
{% elif bmi < 25 %}
<p>Normal</p>
{% elif bmi < 30 %}
<p>Overweight</p>
{% else %}
<p>Obese</p>
{% endif %}
```

Everything is ready, you can now calculate your body mass index. Warning! I am only responsible for bugs; for any other problem, you should consult a nutritionist.

Figure 3.8 – When the form is filled in and Calculate is clicked, the HTML of the component is displayed

Summary

We have acquired the skills to create a bidirectional communication tunnel between a backend and frontend using the WebSockets protocol. We can send plain text, JSON, or HTML—totally asynchronous and without waiting. We even know how to ask the backend to take care of rendering HTML fragments that we will inject without the visitor noticing a delay.

Despite everything we have learned, we still have some issues, such as the fact that the backend can only send information to individual clients, and not to groups. In addition, we still don't know how to interact with the database, create sessions, or identify users. And without all these elements, we will be unable to make an application that facilitates communication between two visitors or manipulate the database. We need to go deeper.

In the next chapter, we will be introduced to database models and create a complete **Browse-Read-Edit-Add-Delete (BREAD)** with a completely new project.

> **BREAD Is an Evolution of CRUD**
>
> **CRUD** is well known when you want to create a complete data processing system (**Create-Read-Update-Delete**). It is traditionally used in interfaces, APIs, databases, and Web Apps, but it does not consider actions such as pagination, search, or sorting. BREAD was born as an expansion to highlight the fact that information must be navigable, browsea: Browse-Read-Edit-Add-Delete.
>
> Reference: `https://guvena.wordpress.com/2017/11/12/bread-is-the-new-crud/`.

4

Working with the Database

This chapter does not aim to teach you how to interact with a database using Django or to create migrations—I assume you already have those minimal skills. This chapter instead will show you how to work with real situations where a Channels instance interacts recurrently with models.

Unless the application is only powered by external APIs, having a database is an elementary requirement in any modern web development. The needs can range from functionality as simple as storing plain text in an orderly fashion, to an authentication system, to managing a complex structure of connections between users. In other words, you must connect to a database if you want to build a practical project.

Fortunately, Django is compatible with the most popular relational databases: PostgreSQL, MariaDB, MySQL, Oracle, and SQLite. And if that's not enough, we can also connect to other possibilities thanks to extensions created by the community: **CockroachDB**, **Firebird**, **Google Cloud Spanner**, and **Microsoft SQL Server**.

We are going to focus on creating a real-time app that manages a database. We will learn how to perform the minimum functionalities of **BREAD: Browse-Read-Edit-Add-Delete**, including simple pagination. And what better example than to create a social network? The information should be saved, sorted, and displayed to all users with as little delay as possible. If there is a very slow response, then we have failed to offer a real-time system and have achieved a boring email system.

For didactic reasons, we will create an anarchist social network. Any visitor, without prior registration, will be able to manipulate the data of any user. If you find it unsafe, you can create a disclaimer page invoking human kindness and suggest not to change other people's content or wait for the next chapters where we will incorporate a complete registration and identification system.

We'll cover the following topics in the chapter:

- Inserting information into the database
- Rendering database information
- Limiting queries with a pager
- Deleting rows from a database
- Updating rows in a database

In addition, we will incorporate some CSS lines to enhance the visual impact, and we will take all the logic to the backend, leaving only the responsibility of managing events on the client side.

Technical requirements

We will build on all the accumulated knowledge from the previous chapters. If you want to start with a template, you can use the following structure, which I will use for future projects:

`https://github.com/PacktPublishing/Building-SPAs-with-Django-and-HTML-Over-the-Wire/tree/main/chapter-4/initial-template`

Inside you will find a sample project that is already prepared with different points that we have touched upon in the previous chapters:

- Integration with Docker
- Minimum structure for working with Channels
- Connection to the database with PostgreSQL
- An HTML home page
- A minimal JavaScript file that connects to Channels

For this project, I created a fork of the template and made some minor changes. You can use either of the two templates, although I recommend the fork for simple aesthetics. You can download it from here: `https://github.com/PacktPublishing/Building-SPAs-with-Django-and-HTML-Over-the-Wire/tree/main/chapter-4/social-network_step_1`. I have changed the project name to `social_network` and the app to `website`. I have also renamed the consumer to `SocialNetworkConsumer`.

Finally, a schema has been added to the database, or model, called `Message` in `app/website/models.py`, which we will use to manage all the messages created by users:

```
from django.db import models
```

```
class Message(models.Model):

    author = models.CharField(max_length=100)
    text = models.TextField(max_length=200)
    created_at = models.DateTimeField(auto_now_add=True)

    class Meta:
        db_table = "messages"
        verbose_name_plural = "Messages"

    def __str__(self):
        return self.text[:10] + "..."
```

The fields included are minimal: author to store the author's name, text for the text of the message, and created_at to have the creation date for later sorting.

To set up the project, you must use Docker:

```
docker-compose up
```

Open your favorite browser to enter http://social-network.localhost. You should find the following result.

Social Network

Figure 4.1 – Displaying a static HTML page on the domain http://social-network.localhost

Visually it is too... minimalistic. But it contains all the elementary pieces to work with.

Next, we will start building the application step by step, touching on the whole flow of editing the Message table or querying it afterward.

Inserting information into the database

We are now ready with an almost empty project but perfectly configured with Channels, rendering a simple static HTML. The first step is to INSERT or save new information in the database. For this, we need a minimum of HTML. We are going to include a form with two fields: name and message. We will also leave a space to show the future messages that we list.

Create an HTML file in app/website/templates/index.html with the following content.

First, we will incorporate a CSS file and a JavaScript file. For the moment, we incorporate the files in the headers:

```
{% load static %}
<!doctype html>
<html lang="en">
<head>
    <meta charset="UTF-8">
    <meta name="viewport"
          content="width=device-width, user-scalable=no,
             initial-scale=1.0, maximum-scale=1.0,
                minimum-scale=1.0">
    <title>Social Network</title>
    <link rel="stylesheet" href="{% static 'css/main.css' %}">
    <script defer src="{% static 'js/index.js' %}">
    </script>
</head>
```

Next, in order to have the host and the scheme (HTTP or HTTPS), we must incorporate it as a dataset that we will later collect in the JavaScript. We have added a form box and another one to list messages, which we will not use for the time being:

```
<body
        data-host="{{ request.get_host }}"
        data-scheme="{{ request.scheme }}"
>
```

The following will be the HTML form we will use to capture and save the new messages. JavaScript will be in charge, in the future, of retrieving all the information:

```html
<div class="container">
    <header>
        <h1>Social Network</h1>
    </header>
    <main id="main">

        <section class="message-form">
            <form>
                <input
                        type="text"
                        placeholder="Name"
                        id="message-form__author"
                        class="message-form__author"
                            class="message-form__author"
                        name="author"
                        required
                >
                <textarea
                        name="message"
                        placeholder="Write your message
                            here..."
                        id="message-form__text"
                        class="message-form__text"
                            class="message-form__text"
                        required
                ></textarea>
                <input
type="submit"
class="message-form__submit"id="message-form__submit"
value="Send"

                >
```

```
        </form>
    </section>
    <!-- End Form for adding new messages -->
```

We'll then define a place to list all messages received from the database. We will also include buttons to paginate the results in blocks of five elements:

```
    <section id="messages">
        <div id="messages__list"></div>
        <button class="button"
            id="messages__previous-page" disabled>
                Previous</button>
        <button class="button" id="messages__next-
            page">Next</button>
    </section>
    <!-- End Messages -->
        </main>
    </div>
</body>
</html>
```

Inside /static/css/main.css, we will add some minimal styles to feel like we are in the 21st century:

```
:root {
    --color__background: #f6f4f3;
    --color__gray: #ccc;
}
```

We add typography that brightens things up for the eye a bit. Helvetica goes well with everything, but you are free to use whatever you like. You won't hurt Django's feelings:

```
* {
    font-family: "Helvetica Neue", Helvetica, Arial, sans-
serif;
    box-sizing: border-box;
}
```

Some of the improvements we added are to correct the margin of the body and to center the content with `container`:

```css
body {
    margin: 0;
    background-color: var(--color__background);
}


.container {
    margin: 0 auto;
    padding: 1rem;
    max-width: 40rem;
}
```

We'll also style the buttons so they don't look like they're from the 90s:

```css
. button {
    display: inline-block;
    padding: 0.5rem 1rem;
    background-color: var(--color__gray);
    border: 0;
    cursor: pointer;
    text-decoration: none;
}


. button:hover {
    filter: brightness(90%);
}
```

And we use a less retro look for forms or elements. Design should not be in conflict with backend work:

```css
.message-form__author, .message-form__text {
    display: block;
    width: 100%;
    outline: none;
    padding: .5rem;
    resize: none;
    border: 1px solid var(--color__gray);
```

```css
    box-sizing: border-box;
}

.message-form__submit {
    display: block;
    width: 100%;
    outline: none;
    padding: .5rem;
    background-color: var(--color__gray);
    border: none;
    cursor: pointer;
    font-weight: bold;
}

.message-form__submit:hover {
    filter: brightness(90%);
}

.message {
    border: 1px solid var(--color__gray);
    border-top: 0;
    padding: .5rem;
    border-radius: .5rem;
}

.message__author {
    font-size: 1rem;
}

.message__created_at {
    colour: var(--color__gray);
}

.message__footer {
    display: flex;
    justify-content: space-between;
```

```
        align-items: center;
}
```

If you pull up Docker, and you enter `http://social-network.localhost`, you will find the following page:

Social Network

| Name |
| Write your message here... |
| **Send** |

Figure 4.2 – The form for entering a new message with CSS styles

Feel free to add whatever you need, and even a CSS framework.

Now we are going to include the JavaScript events to send the form data to the consumer. We will create a new file in `/static/js/index.js` with the following content:

```
/*
    VARIABLES
*/
// Connect to WebSockets server (SocialNetworkConsumer)
const myWebSocket = new WebSocket(`${document.body.
    dataset.scheme === 'http' ? 'ws' : 'wss'}://${
    document.body.dataset.host }/ws/social-network/`);
const inputAuthor = document.querySelector("#message-
    form__author");
const inputText = document.querySelector("#message-
    form__text");
const inputSubmit = document.querySelector("#message-
    form__submit");

/*
    FUNCTIONS
*/
```

```
/**
 * Send data to WebSockets server
 * @param {string} message
 * @param {WebSocket} webSocket
 * @return {void}
 */
function sendData(message, webSocket) {
    webSocket.send(JSON.stringify(message));
}



/**
 * Send new message
 * @param {Event} event
 * @return {void}
 */
function sendNewMessage(event) {
    event.preventDefault();
    // Prepare the information we will send
    const newData = {
        "action": "add message",
        "data": {
            "author": inputAuthor.value,
            "text": inputText.value
        }
    };
    // Send the data to the server
    sendData(newData, myWebSocket);
    // Clear message form
    inputText.value = "";
}

/*
    EVENTS
*/
```

```javascript
// Event when a new message is received by WebSockets
myWebSocket.addEventListener("message", (event) => {
    // Parse the data received
    const data = JSON.parse(event.data);
    // Renders the HTML received from the Consumer
    document.querySelector(data.selector). innerHTML =
        data.html;
});


// Sends new message when you click on Submit
inputSubmit.addEventListener("click", sendNewMessage);
```

In the variable section, we capture all the HTML elements that we need to capture all events and create a WebSockets connection. The sendData function is used to send messages to the backend, and sendNewMessage is executed when we click on the **Submit** button. The JSON will always be sent with the following structure:

```json
{
        "action": "text",
        "data": {}
}
```

We modify the consumer to receive the information and save it. Edit app/website/consumers.py with the following content:

```python
from channels.generic.websocket import JsonWebsocketConsumer
from django.template.loader import render_to_string
from . models import Message
from asgiref.sync import async_to_sync


class SocialNetworkConsumer(JsonWebsocketConsumer):

    room_name = 'broadcast

    def connect(self):
        """Event when client connects"""
        # Accept the connection
```

```python
        self.accept()
        # Assign the Broadcast group
        async_to_sync(self.channel_layer.group_add)
            (self.room_name, self.channel_name)
        # Send you all the messages stored in the database.

    def disconnect(self, close_code):
        """Event when client disconnects"""
        # Remove from the Broadcast group
        async_to_sync(self.channel_layer.group_discard)
            (self.room_name, self.channel_name)

    def receive_json(self, data_received):
        """
            Event when data is received
            All information will arrive in 2 variables:
            'action', with the action to be taken
            'data' with the information
        """

        # Get the data
        data = data_received['data']
        # Depending on the action we will do one task or
        another.
        match data_received['action']:
            case 'add message':
                # Add message to database
                Message.objects.create(
                    author=data['author'],
                    text=data['text'],
                )
```

You can see that the client is added to a room when connecting and removed when disconnecting. In *Chapter 5, Separating Communication in Rooms*, we will talk in depth about the possibilities of rooms, but for now, we will group all users in a single room with the name *broadcast*.

When an action with the text `'add message'` is received, we directly create a new message with the information we are sending from the frontend.

We already stored information! Although we can't see it or sort it at the moment.

All the code up to this point can be found in the following repository, which is the first part of the activity:

https://github.com/PacktPublishing/Building-SPAs-with-Django-and-HTML-Over-the-Wire/tree/main/chapter-4/social-network_step_2

In the next section, we will print all the messages we have saved directly in HTML via WebSockets.

Rendering database information

We have already built a form that sends a new message to the backend via a WebSockets connection, which we capture in the consumer and then store in the database.

Now we're going to get all the messages from the database, render them with Django's template engine, send the HTML to the client over the connection we kept, and inject the content into the appropriate frontend element.

Create the HTML template that will generate all the message blocks in the path /app/website/templates/components/_list-messages.html with the content:

```
{% for message in messages %}
    <article class="message" id="message--{{ message.id
        }}">
        <h2 class="message__author">{{ message.author }}
        </h2>
        <p class="message__text">{{ message.text }}</p>
        <footer class="message__footer">
            <time class="message__created_at">{{
                message.created_at }}</time>
            <div class="message__controls">
                <button class="button messages__update"
                  data-id="{{ message.id }}"> Edit</button>
                <button class="button messages__delete"
                    data-id="{{ message.id }}"> Delete
                    </button>
            </div>
        </footer>
    </article>
{% endfor %}
```

At the moment, the **Edit** and **Delete** buttons are for decoration. Later, we will give them their functionality.

Edit the consumer, which is in `app/website/consumers.py`, to include an action that returns messages:

```python
from channels.generic.websocket import JsonWebsocketConsumer
from django.template.loader import render_to_string
from . models import Message
from asgiref.sync import async_to_sync


class SocialNetworkConsumer(JsonWebsocketConsumer):

    room_name = 'broadcast'

    def connect(self):
        """Event when client connects"""
        # Accept the connection
        self.accept()
        # Assign the Broadcast group
        async_to_sync(self.channel_layer.group_add)
            (self.room_name, self.channel_name)
        # Send you all the messages stored in the database.
        self.send_list_messages()

    def disconnect(self, close_code):
        """Event when client disconnects"""
        # Remove from the Broadcast group
        async_to_sync(self.channel_layer.group_discard)
            (self.room_name, self.channel_name)

    def receive_json(self, data_received):
        """
        Event when data is received
        All information will arrive in 2 variables:
        'action', with the action to be taken
        'data' with the information
        """
```

```python
        # Get the data
        data = data_received['data']
        # Depending on the action we will do one task or
          another.
        match data_received['action']:
            case 'add message':
                # Add message to database
                Message.objects.create(
                    author=data['author'],
                    text=data['text'],
                )
                # Send messages to all clients
                self.send_list_messages()
            case 'list messages':
                # Send messages to all clients
                self.send_list_messages()


    def send_html(self, event):
        """Event: Send html to client"""
        data = {
            'selector': event['selector'],
            'html': event['html'],
        }
        self.send_json(data)


    def send_list_messages(self):
        """ Send list of messages to client"""
        # Filter messages to the current page
        messages = Message.objects.order_by('-created_at')
        # Render HTML and send to client
        async_to_sync(self.channel_layer.group_send)(
            self.room_name, {
                type': 'send.html', # Run 'send_html()'
```

```
                        method
                'selector': '#messages__list',
                'html': render_to_string
                ('components/_list-messages.html', {
                    'messages': messages})
            }
        )
```

When the frontend sends us the `'list messages'` action or creates a WebSockets connection, we will execute the `send_list_messages()` function. The backend will get all the messages, render the HTML, and send the messages to the frontend. Inside the function, we are performing a query to get all the messages in descending order, an elementary action if you have worked with models before. The important thing happens in the next snippet:

```
async_to_sync(self.channel_layer.group_send)(
        self.room_name, {
            type': 'send.html', # Run 'send_html()'
                method
            'selector': '#messages__list',
            'html': render_to_string('components/_list-
                messages.html', { 'messages': messages})
        }
    )
```

When sending information to a group, it will always be done asynchronously so as not to block the main thread, but when communicating with the database, it must be synchronous. How can both types of logic coexist? By turning `self.channel_layer.group_send()` into a synchronous function thanks to `async_to_sync()`.

`self.channel_layer.group_send()` is an unusual function. Its first argument must be the name of the room where you want to send the information, which in our case will be `self.room_name`, which is declared at the beginning of the consumer. And the second argument must have a dictionary where `type` is the name of the function to execute (if you have `_`, it must be replaced with a dot), and the rest of the keys are the information that we want to pass to the function. Inside `send_html`, we capture the previous information with `event[]`. And finally, we send the data to the client in the same way as in *Chapter 3, Adding WebSockets to Django*, with `send_json()`.

When we want to inject HTML, we will send the `html` key with the rendered HTML and `selector` to tell JavaScript where to inject it. The backend will decide what and where each element should go.

When viewing the browser, we will find the messages that we have been adding, as shown in *Figure 4.3* :

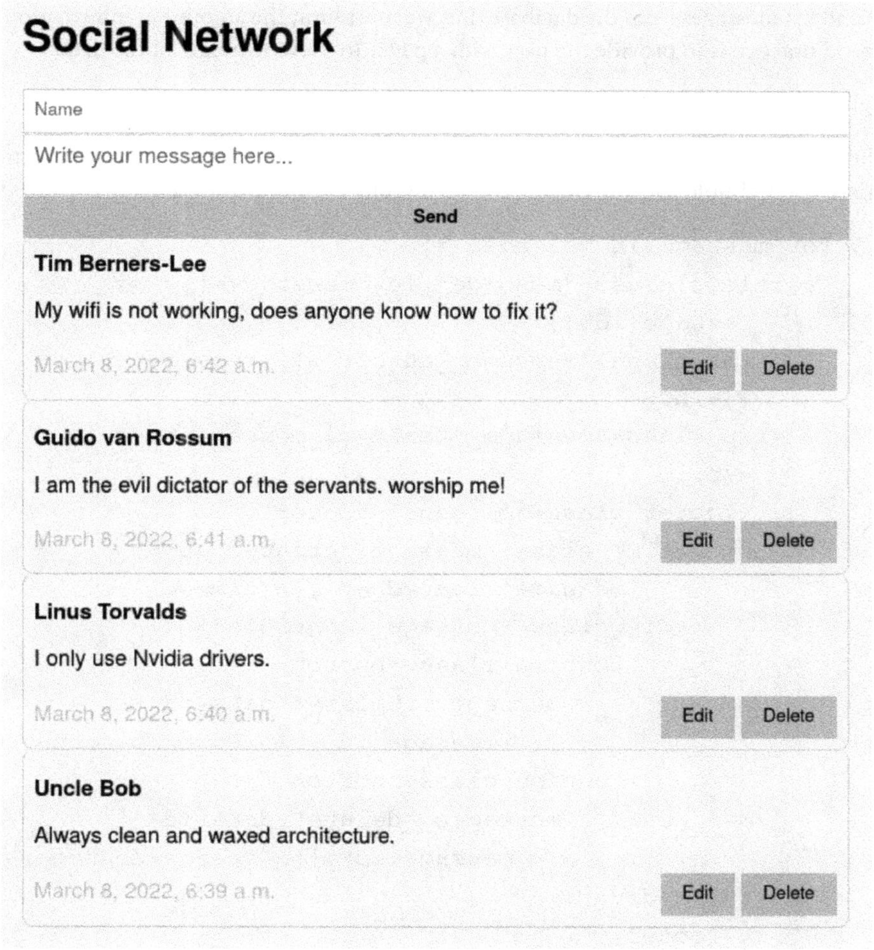

Figure 4.3 – All messages we have saved in the database are displayed

All the code up to this point can be found at the following link:

```
https://github.com/PacktPublishing/Building-SPAs-with-Django-and-
HTML-Over-the-Wire/tree/main/chapter-4/social-network_step_3
```

What would happen if we were to display hundreds or thousands of messages per user? The performance and memory consequences would be catastrophic. We must avoid fireworks by limiting the number of messages that can be displayed. Therefore, we will limit it to five messages per page and add some buttons to navigate between each slice. Let's see that in the next section.

Limiting queries with a pager

We can add and list messages from the database. But we must limit the amount of information a user can see. A good practice is to provide the user with a pager to move through all the data.

Follow the given steps to add a simple pager:

1. Modify the template `/app/website/templates/components/_list-messages.html` to add a simple pager divided into two buttons (forward and back):

```
{% for message in messages %}
    <article class="message" id="message--{{
        message.id }}">
        <h2 class="message__author">{{ message.author
        }}</h2>
        <p class="message__text">{{ message.text }}
        </p>
        <footer class="message__footer">
            <time class="message__created_at">{{
                message.created_at }}</time>
            <div class="message__controls">
                <button class="button
                    messages__update" data-id="{{
                        message.id }}"> Edit</button>
                <button class="button
                    messages__delete" data-id="{{
                        message.id }}"> Delete</button>
            </div>
        </footer>
    </article>
{% endfor %}
{% if total_pages != 0 %}
    <!-- Paginator -->
    <div id="paginator" data-page="{{ page }}">
        {# The back button on the first page is not
        displayed #}
        {% if page ! = 1 %}
        <button class="button" id="messages__previous-
            page"> Previous</button>
```

```
            {% endif %}

            {# The forward button on the last page is not
            displayed #}
            {% if page ! = total_pages %}
            <button class="button" id="messages__next-
                page">Next</button>
            {% endif %}
        </div>
        <!-- End Paginator -->
    {% endif %}
```

- With `data-page="{{ page }}"`, we are giving JavaScript a counter with the page we are on. We will use this data to create a new event that will trigger an action indicating whether we want to go to the next page or back.

- The conditional `{% if page ! = 1 %}` is used to avoid showing the back button if we are on the first page.

- The conditional `{% if page ! = total_pages %}` ignores the rendering of the forward button if we are on the last page.

2. We add to the consumer (`/app/website/consumers.py`) a slice system for `send_list_messages()`:

```python
from channels.generic.websocket import
JsonWebsocketConsumer
from django.template.loader import render_to_string
from . models import Message
from asgiref.sync import async_to_sync
import math


class SocialNetworkConsumer(JsonWebsocketConsumer):

    room_name = 'broadcast
    max_messages_per_page = 5 # New line

    def connect(self):
```

```python
        """Event when client connects"""
        # Accept the connection
        self.accept()
        # Assign the Broadcast group
        async_to_sync(self.channel_layer.group_add)(self.
room_name, self.channel_name)
        # Send you all the messages stored in the
database.
        self.send_list_messages()

    def disconnect(self, close_code):
        """Event when client disconnects"""
        # Remove from the Broadcast group
        async_to_sync(self.channel_layer.
            group_discard)(self.room_name,
                self.channel_name)

    def receive_json(self, data_received):
        """
        Event when data is received
        All information will arrive in 2
        variables:
        'action', with the action to be taken
        'data' with the information
        """

        # Get the data
        data = data_received['data']
        # Depending on the action we will do one task
         or another.
        match data_received['action']:
            case 'add message':
                # Add message to database
                Message.objects.create(
                    author=data['author'],
                    text=data['text'],
```

```python
            )
            # Send messages to all clients
            self.send_list_messages()
        case 'list messages':
            # Send messages to all clients
            self.send_list_messages(data['page'])
            # Update line


def send_html(self, event):
    """Event: Send html to client"""
    data = {
        'selector': event['selector'],
        'html': event['html'],
    }
    self.send_json(data)


def send_list_messages(self, page=1):
# Update line
    "Send list of messages to client"""""
    # Filter messages to the current page
    start_pager = self.max_messages_per_page * \
        (page - 1) # New line
    end_pager = start_pager + \
        self.max_messages_per_page # New line
    messages = Message.objects.order_by('-\
        created_at')
    messages_page = messages[start_pager:\
        end_pager] # New line
    # Render HTML and send to client
    total_pages = math.ceil(messages.count() / \
        self.max_messages_per_page) # New line
    async_to_sync(self.channel_layer.group_send)(
        self.room_name, {
            'type': 'send.html', # Run
```

```
                    'send_html()' method
                'selector': '#messages__list',
                'html': render_to_string
                  ('components/_list-messages.html', {
                    'messages': messages_page,
                    # Update line
                    'page': page, # New line
                    total_pages': total_pages,
                    # New line
                })
            }
        )
```

- The variable max_messages_per_page = 5 has been added to indicate the number of items per page.

- The 'list messages' action now collects and passes to the send_list_messages function the page to be displayed.

- We have updated send_list_messages. We allow you to indicate the page to be displayed, we calculate the query slices, and we give render_to_string the messages with the slice, page, and total_pages variables, which we use to know if we are on the last page.

3. In /static/js/index.js, add the two JavaScript functions (goToNextPage and goToPreviousPage) that will take care of the page-turning. Actually, they just use an action with the request to list the messages but in another slice:

```
/*
    VARIABLES
*/
// Connect to WebSockets server (SocialNetworkConsumer)
const myWebSocket = new WebSocket
(`${document.body.dataset.scheme === 'http' ? 'ws' :
'wss'}://${ document.body.dataset.host }/ws/social-
    network/`);
const inputAuthor = document.querySelector("#message-
    form__author");
const inputText = document.querySelector("#message-
    form__text");
```

```javascript
const inputSubmit = document.querySelector("#message-
    form__submit");


/*

    FUNCTIONS
*/


/**
 * Send data to WebSockets server
 * @param {string} message
 * @param {WebSocket} webSocket
 * @return {void}
 */
function sendData(message, webSocket) {
    webSocket.send(JSON.stringify(message));
}



/**
 * Send new message
 * @param {Event} event
 * @return {void}
 */
function sendNewMessage(event) {
    event.preventDefault();
    // Prepare the information we will send
    const newData = {
        "action": "add message",
        "data": {
            "author": inputAuthor.value,
            "text": inputText.value
        }
    };
    // Send the data to the server
    sendData(newData, myWebSocket);
    // Clear message form
```

```
        inputText.value = "";
}


/**
 * Get current page stored in #paginator as dataset
 * @returns {number}
 */
function getCurrentPage() {
    return parseInt(document.
        querySelector("#paginator"). dataset.page);
}


/**
 * Switch to the next page
 * @param {Event} event
 * @return {void}
 */
function goToNextPage(event) {
    // Prepare the information we will send
    const newData = {
        "action": "list messages",
        "data": {
            "page": getCurrentPage() + 1,
        }
    };
    // Send the data to the server
    sendData(newData, myWebSocket);
}


/**
 * Switch to the previous page
 * @param {Event} event
 * @return {void}
 */
function goToPreviousPage(event) {
```

```javascript
        // Prepare the information we will send
        const newData = {
            "action": "list messages",
            "data": {
                "page": getCurrentPage() - 1,
            }
        };
        // Send the data to the server
        sendData(newData, myWebSocket);
}


/*

    EVENTS
*/


// Event when a new message is received by WebSockets
myWebSocket.addEventListener("message", (event) => {
    // Parse the data received
    const data = JSON.parse(event.data);
    // Renders the HTML received from the Consumer
    document.querySelector(data.selector). innerHTML =
        data.html;
    /* Reassigns the events of the newly rendered HTML */
    // Pagination
    document.querySelector("#messages__next-page")?.
        addEventListener("click", goToNextPage);
    document.querySelector("#messages__previous-
        page")?. addEventListener("click",
            goToPreviousPage);
});


// Sends new message when you click on Submit
inputSubmit.addEventListener("click", sendNewMessage);
```

And where do we add the listeners for the buttons to execute the functions? Inside the "message" event of WebSockets. Why? Every time the message section is updated, all the HTML is deleted and recreated with what is received from the backend. The events disappear with every update. We have to reassign them. After redrawing the messages, we will assign the listeners:

```
document.querySelector("#messages__next-page")?.
    addEventListener("click", goToNextPage);
document.querySelector("#messages__previous-page")?.
    addEventListener("click", goToPreviousPage);
```

Go ahead, create as many messages as you want – at least six – to check how the pager does its job properly.

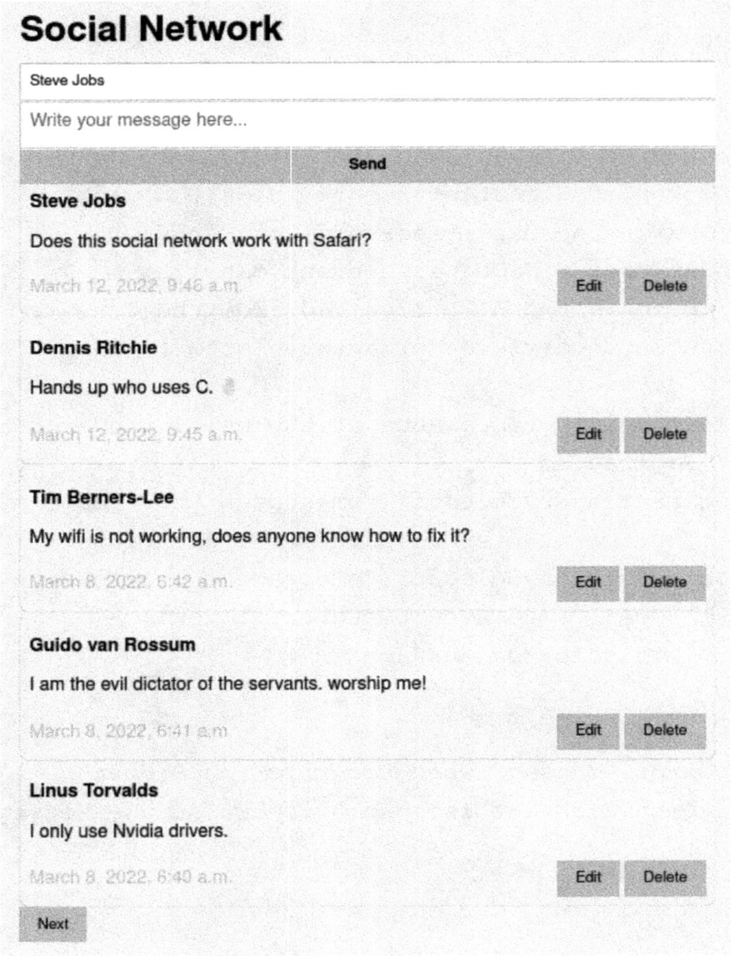

Figure 4.4 – We display the first page of messages

If we turn the page by clicking on the **Next** button, the next block of messages will be rendered.

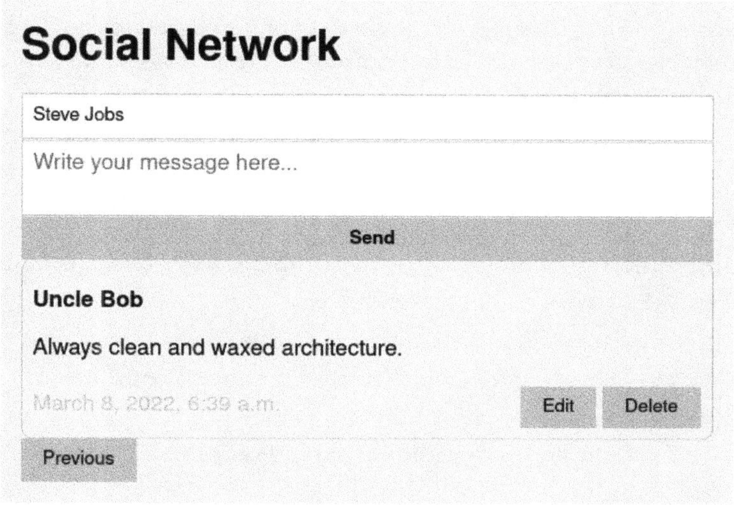

Figure 4.5 – We display the last page of messages

All the code up to this point can be found at the following link:

`https://github.com/PacktPublishing/Building-SPAs-with-Django-and-HTML-Over-the-Wire/tree/main/chapter-4/social-network_step_4`

The next target will be to delete messages with the **Delete** button. In the next section, we will send via `WebSockets` an action with the instruction to delete messages with the concrete `id` of a message—we will be as precise as a sniper.

Deleting rows from a database

In the previous sections, we managed to build a system where we could add new messages, list them, and paginate them. But so far, we are unable to delete anything.

The way the project is structured, it is really quick to implement:

1. We check in `/app/website/templates/components/_list-messages.html` that we are adding to each button a dataset with the `id`. We do this task when we list the messages; we must be aware of the source where the `id` that we will use comes from to know which message we must delete:

    ```
    <button class="button messages__delete" data-id="{{
        message.id }}"> Delete</button>
    ```

2.   In /static/js/index.js, add the deleteMessage function. We will capture the
     dataset with the id and send it to the consumer with the action "delete message". In
     addition, we will add each listener after the listener of the pager. Let's not forget the reason
     for this positioning: all events must be reassigned after every change or new backend message
     with HTML that we inject:

```
/*

    VARIABLES
*/
// Connect to WebSockets server (SocialNetworkConsumer)
const myWebSocket = new WebSocket
    (`${document.body.dataset.scheme === 'http' ? 'ws'
    : 'wss'}://${ document.body.dataset.host
    }/ws/social-network/`);
const inputAuthor = document.querySelector("#message-
    form__author");
const inputText = document.querySelector("#message-
    form__text");
const inputSubmit = document.querySelector("#message-
    form__submit");

/*

    FUNCTIONS
*/

/**
 * Send data to WebSockets server
 * @param {string} message
 * @param {WebSocket} webSocket
 * @return {void}
 */
function sendData(message, webSocket) {
    webSocket.send(JSON.stringify(message));
}

/**
 * Delete message
```

```
 * @param {Event} event
 * @return {void}
 */
function deleteMessage(event) {
    const message = {
        "action": "delete message",
        "data": {
            "id": event.target.dataset.id
        }
    };
    sendData(message, myWebSocket);
}


/**
 * Send new message
 * @param {Event} event
 * @return {void}
 */
function sendNewMessage(event) {
    event.preventDefault();
    // Prepare the information we will send
    const newData = {
        "action": "add message",
        "data": {
            "author": inputAuthor.value,
            "text": inputText.value
        }
    };
    // Send the data to the server
    sendData(newData, myWebSocket);
    // Clear message form
    inputText.value = "";
}


/**
```

```
 * Get current page stored in #paginator as dataset
 * @returns {number}
 */
function getCurrentPage() {
    return parseInt(document.querySelector("#paginator").
dataset.page);
}



/**
 * Switch to the next page
 * @param {Event} event
 * @return {void}
 */
function goToNextPage(event) {
    // Prepare the information we will send
    const newData = {
        "action": "list messages",
        "data": {
            "page": getCurrentPage() + 1,
        }
    };
    // Send the data to the server
    sendData(newData, myWebSocket);
}

/**
 * Switch to the previous page
 * @param {Event} event
 * @return {void}
 */
function goToPreviousPage(event) {
    // Prepare the information we will send
    const newData = {
        "action": "list messages",
        "data": {
```

```
                    "page": getCurrentPage() - 1,
            }
    };
    // Send the data to the server
    sendData(newData, myWebSocket);
}


/*

    EVENTS
*/


// Event when a new message is received by WebSockets
myWebSocket.addEventListener("message", (event) => {
    // Parse the data received
    const data = JSON.parse(event.data);
    // Renders the HTML received from the Consumer
    document.querySelector(data.selector). innerHTML =
        data.html;
    /* Reassigns the events of the newly rendered HTML */
    // Pagination
    document.querySelector("#messages__next-
      page")?.addEventListener("click", goToNextPage);
    document.querySelector("#messages__previous-
        page")?.addEventListener("click",
            goToPreviousPage);
    // Add to all delete buttons the event
    document.querySelectorAll
        (".messages__delete").forEach(button => {
        button.addEventListener("click",
            deleteMessage);
    });
});


// Sends new message when you click on Submit
inputSubmit.addEventListener("click", sendNewMessage);
```

3. Now, edit /app/website/consumers.py with the action 'delete message':

```python
# app/website/consumers.py
from channels.generic.websocket import
JsonWebsocketConsumer
from django.template.loader import render_to_string
from .models import Message
from asgiref.sync import async_to_sync
import math


class SocialNetworkConsumer(JsonWebsocketConsumer):

    room_name = 'broadcast'
    max_messages_per_page = 5

    def connect(self):
        """Event when client connects"""
        # Accept the connection
        self.accept()
        # Assign the Broadcast group
        async_to_sync(self.channel_layer.group_add)
            (self.room_name, self.channel_name)
        # Send you all the messages stored in the
        database.
        self.send_list_messages()

    def disconnect(self, close_code):
        """Event when client disconnects"""
        # Remove from the Broadcast group
        async_to_sync(self.channel_layer.group_
            discard)(self.room_name, self.channel_name)

    def receive_json(self, data_received):
        """
            Event when data is received
            All information will arrive in 2 variables:
```

```
                    'action', with the action to be taken
                    'data' with the information
            """

            # Get the data
            data = data_received['data']
            # Depending on the action we will do one task or
another.
            match data_received['action']:
                case 'add message':
                    # Add message to database
                    Message.objects.create(
                        author=data['author'],
                        text=data['text'],
                    )
                    # Send messages to all clients
                    self.send_list_messages()
                case 'list messages':
                    # Send messages to all clients
                    self.send_list_messages(data['page'])
                case 'delete message':
                    # Delete message from database
                    Message.objects.get
                        (id=data['id']).delete() # New line
                    # Send messages to all clients
                    self.send_list_messages() # New line


    def send_html(self, event):
        """Event: Send html to client"""
        data = {
            'selector': event['selector'],
            'html': event['html'],
        }
        self.send_json(data)
```

```python
def send_list_messages(self, page=1):
    """Send list of messages to client"""
    # Filter messages to the current page
    start_pager = self.max_messages_per_page *
        (page - 1)
    end_pager = start_pager +
        self.max_messages_per_page
    messages = Message.objects.order_by('-
        created_at')
    messages_page = messages
        [start_pager:end_pager]
    # Render HTML and send to client
    total_pages = math.ceil(messages.count() /
        self.max_messages_per_page)
    async_to_sync(self.channel_layer.group_send)(
        self.room_name, {
            'type': 'send.html', # Run
            'send_html()' method
            'selector': '#messages__list',
            'html': render_to_string
              ('components/_list-messages.html', {
                'messages': messages_page,
                'page': page,
                'total_pages': total_pages,
            })
        }
    )
```

With `Message.objects.get(id=data['id']).delete()`, we delete the message directly from the `id` sent to us by the frontend. Finally, we update the list of messages from all clients with `self.send_list_messages()`.

All the code up to this point can be found at the following link:

`https://github.com/PacktPublishing/Building-SPAs-with-Django-and-HTML-Over-the-Wire/tree/main/chapter-4/social-network_step_5`

We have finished creating the functionality to delete rows in the database. We are now able to selectively delete messages. In the next part of the exercise, we will finish building the social network by adding the ability to modify an existing message. And with this new feature, we will have all the letters of BREAD.

Updating rows in a database

In this last part of the exercise, we will finish building the social network by adding a form to modify the information.

All the finished code can be found at the following link:

```
https://github.com/PacktPublishing/Building-SPAs-with-Django-and-
HTML-Over-the-Wire/tree/main/chapter-4/social-network_step_6
```

Let's start:

1. Create a new HTML component, in /app/website/templates/components/_edit-message.html, with the following content:

    ```
    <form class="update-form" data-id="{{ message.id }}">
        <input
                type="text"
                placeholder="Name"
                id="message-form__author--update"
                class="message-form__author"
                name="author"
                required
                value="{{ message.author }}"
        >
        <textarea
                name="message"
                placeholder="Write your message here..."
                id="message-form__text--update"
                class="message-form__text"
                required
        >{{ message.text }}</textarea>
        <input
                type="submit"
                class="message-form__submit" class="message-
    form__submit"
    ```

```
                    id="message-form__submit--update"
                    value="Update"
        >
    </form>
```

The HTML component is practically the same as when we create a message, except for the detail that we store the id of the message to be modified in a dataset that we host in the `<form>` tag with `data-id="{{ message.id }}"` and that we fill in all the fields.

2. Create an action requesting the edit form in the consumer:

```
case 'open edit page':
self.open_edit_page(data['id'])
```

This action will only render and send the previous component so that the user can edit the following information:

```
def open_edit_page(self, id):
        """Send the form to edit the message"""
        message = Message.objects.get(id=id)
        async_to_sync(self.channel_layer.group_send)(
            self.room_name, {
                'type': 'send.html', # Run
                    'send_html()' method
                'selector': f'#message--{id}',
                'html': render_to_string
                    ('components/_edit-message.html',
                        {'message': message})
            }
        )
```

3. Now, in the consumer, add the action to collect the information from the form, update the database, and render the list of messages:

```
case 'update message':
                # Update message in database
                Message.objects.filter(id=data['id']).
update(
                        author=data['author'],
                        text=data['text'],
                )
```

```
            # Send messages to all clients
            self.send_list_messages()
```

The entire integrated consumer, including the update action, will look like this:

```python
from channels.generic.websocket import
JsonWebsocketConsumer
from django.template.loader import render_to_string
from .models import Message
from asgiref.sync import async_to_sync
import math


class SocialNetworkConsumer(JsonWebsocketConsumer):

    room_name = 'broadcast'
    max_messages_per_page = 5

    def connect(self):
        """Event when client connects"""
        # Accept the connection
        self.accept()
        # Assign the Broadcast group
        async_to_sync(self.channel_layer.group_add)(self.
room_name, self.channel_name)
        # Send you all the messages stored in the
database.
        self.send_list_messages()

    def disconnect(self, close_code):
        """Event when client disconnects"""
        # Remove from the Broadcast group
        async_to_sync(self.channel_layer.
            group_discard)(self.room_name,
                self.channel_name)

    def receive_json(self, data_received):
```

```python
        """
            Event when data is received
            All information will arrive in 2 variables:
            'action', with the action to be taken
            'data' with the information
        """

        # Get the data
        data = data_received['data']
        # Depending on the action we will do one task or
another.
        match data_received['action']:
            case 'add message':
                # Add message to database
                Message.objects.create(
                    author=data['author'],
                    text=data['text'],
                )
                # Send messages to all clients
                self.send_list_messages()
            case 'list messages':
                # Send messages to all clients
                self.send_list_messages(data['page'])
            case 'delete message':
                # Delete message from database
                Message.objects.get
                    (id=data['id']).delete()
                # Send messages to all clients
                self.send_list_messages()
            case 'open edit page':
                self.open_edit_page(data['id'])
            case 'update message':
                # Update message in database
                Message.objects.filter(id=data['id']).
                    update(
                    author=data['author'],
```

```
                text=data['text'],
            ) # New block
            # Send messages to all clients
            self.send_list_messages() # New line


    def send_html(self, event):
        """Event: Send html to client"""
        data = {
            'selector': event['selector'],
            'html': event['html'],
        }
        self.send_json(data)


    def send_list_messages(self, page=1):
        """Send list of messages to client"""
        # Filter messages to the current page
        start_pager = self.max_messages_per_page * (page
- 1)
        end_pager = start_pager +
            self.max_messages_per_page
        messages = Message.objects.order_by('-
            created_at')
        messages_page =
            messages[start_pager:end_pager].
        # Render HTML and send to client
        total_pages = math.ceil(messages.count() /
            self.max_messages_per_page)
        async_to_sync(self.channel_layer.group_send)(
            self.room_name, {
                'type': 'send.html', # Run
                    'send_html()' method
                'selector': '#messages__list',
                'html': render_to_string
                    ('components/_list-messages.html', {
```

```python
                    'messages': messages_page,
                    'page': page,
                    'total_pages': total_pages,
                })
            }
        )


    def open_edit_page(self, id):
        """Send the form to edit the message"""
        message = Message.objects.get(id=id)
        async_to_sync(self.channel_layer.group_send)(
            self.room_name, {
                'type': 'send.html', # Run
                    'send_html()' method
                'selector': f'#message--{id}',
                'html': render_to_string
                    ('components/_edit-message.html',
                        {'message': message})
            }
        )
```

4. We create the necessary events in the frontend to request the form, collect the information, and send it.

    We connect to the channel and collect the form fields where users can write new messages:

```javascript
/*
    VARIABLES
*/
// Connect to WebSockets server (SocialNetworkConsumer)
const myWebSocket = new WebSocket(`${document.body.
dataset.scheme === 'http' ? 'ws' : 'wss'}://${ document.
body.dataset.host }/ws/social-network/`);
const inputAuthor = document.querySelector("#message-
    form__author");
const inputText = document.querySelector("#message-
    form__text");
```

```
const inputSubmit = document.querySelector("#message-
    form__submit");
```

We use minimal and essential functions such as sending new information, showing the form to update, sending information to update, deleting a specific element, and pagination management:

```
/*

    FUNCTIONS

*/


/**
 * Send data to WebSockets server
 * @param {string} message
 * @param {WebSocket} webSocket
 * @return {void}
 */
function sendData(message, webSocket) {
    webSocket.send(JSON.stringify(message));
}


/**
 * Displays the update form
 * @param {Event} event
 * @return {void}
 */
function displayUpdateForm(event) {
    const message = {
        "action": "open edit page",
        "data": {
            "id": event.target.dataset.id
        }
    };
    sendData(message, myWebSocket);
}


/**
```

```
 * Update message
 * @param {Event} event
 * @return {void}
 */
function updateMessage(event) {
    event.preventDefault();
    const message = {
        "action": "update message",
        "data": {
            "id": event.target.dataset.id,
            "author": event.target.
querySelector("#message-form__author--update"). value,
            "text": event.target.querySelector("#message-
form__text--update"). value
        }
    };
    sendData(message, myWebSocket);
}

/**
 * Delete message
 * @param {Event} event
 * @return {void}
 */
function deleteMessage(event) {
    const message = {
        "action": "delete message",
        "data": {
            "id": event.target.dataset.id
        }
    };
    sendData(message, myWebSocket);
}

/**
 * Send new message
```

```
 * @param {Event} event
 * @return {void}
 */
function sendNewMessage(event) {
    event.preventDefault();
    // Prepare the information we will send
    const newData = {
        "action": "add message",
        "data": {
            "author": inputAuthor.value,
            "text": inputText.value
        }
    };
    // Send the data to the server
    sendData(newData, myWebSocket);
    // Clear message form
    inputText.value = "";
}


/**
 * Get current page stored in #paginator as dataset
 * @returns {number}
 */

function getCurrentPage() {
    return parseInt(document.querySelector("#paginator").
dataset.page);
}


/**
 * Switch to the next page
 * @param {Event} event
 * @return {void}
 */
```

```
function goToNextPage(event) {
    // Prepare the information we will send
    const newData = {
        "action": "list messages",
        "data": {
            "page": getCurrentPage() + 1,
        }
    };
    // Send the data to the server
    sendData(newData, myWebSocket);
}


/**
 * Switch to the previous page
 * @param {Event} event
 * @return {void}
 */
function goToPreviousPage(event) {
    // Prepare the information we will send
    const newData = {
        "action": "list messages",
        "data": {
            "page": getCurrentPage() - 1,
        }
    };
    // Send the data to the server
    sendData(newData, myWebSocket);
}
```

The most important event that receives information from the backend is "message".
Every time we receive new data, we print it and recapture all events. Without this constant
re-assignment, we would lose all events on every new rendering or redrawing of HTML:

```
/*
    EVENTS
*/
```

```javascript
// Event when a new message is received by WebSockets
myWebSocket.addEventListener("message", (event) => {
    // Parse the data received
    const data = JSON.parse(event.data);
    // Renders the HTML received from the Consumer
    document.querySelector(data.selector).innerHTML =
        data.html;
    /* Reassigns the events of the newly rendered HTML */
    // Pagination
    document.querySelector("#messages__next-page")?.
        addEventListener("click", goToNextPage);
    document.querySelector("#messages__previous-
        page")?.addEventListener("click",
            goToPreviousPage);
    // Add to all delete buttons the event
    document.querySelectorAll(". messages__delete").
forEach(button => {
        button.addEventListener("click", deleteMessage);
    });
    // Add to all update buttons the event
    document.querySelectorAll(". messages__update").
forEach(button => {
        button.addEventListener("click",
displayUpdateForm);
    });
    // Add to the update form the event
    document.querySelectorAll(". update-form").
forEach(form => {
        form.addEventListener("submit", updateMessage);
    });
});

// Sends new message when you click on Submit
inputSubmit.addEventListener("click", sendNewMessage);
```

All of the preceding code is the final version of JavaScript:

- The function `displayUpdateForm` has been added to ask the consumer to draw the edit form in the same place where the message is located.

- The `updateMessage` function has been created to send new information to the consumer in order to update the message.

- Button listeners are included for updating right after the paging and deleting events.

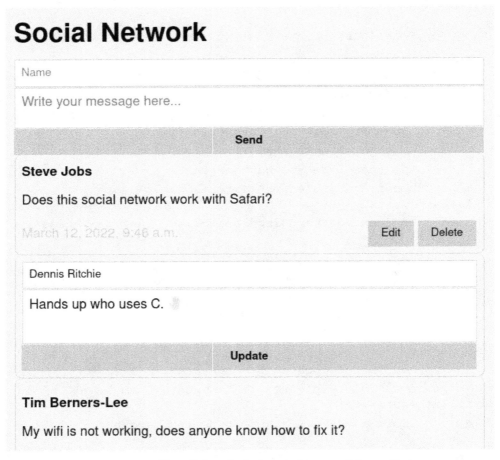

Figure 4.6 – The edit form is displayed when you click on edit

We did it! The BREAD is complete. We can now spread it with butter and let it be consumed by as many customers as possible.

Remember to open the exercise with different browsers to appreciate the magic of synchronization. Every action of any user will be visualized by the rest.

Summary

We have been able to connect a consumer to a database to manage its information and reply with new rendered HTML structures that we have injected. Gradually, a very basic real-time social network has been built to insert messages, list them, filter them, update them, and delete them. The frontend has a simple role – handling events, sending data, and receiving HTML via WebSockets.

Currently, there are several limitations related to group discrimination. When an action is performed, it is propagated to all users, meaning all actions have an impact on all visitors at the same time. Basically, that's a good thing that we want it to happen, but not in all flows. Do I want everyone to update their message listings when a new message is inserted? Yes, of course – and when editing or deleting. Although it should be avoided for certain actions that should be private. At the moment, if one user changes page, everyone changes page. That's why we are going to delve into the possibilities offered by the Channels Rooms: a mechanism that allows us to send data to a specific client or a group defined by us. With this technique, we can go a step further by incorporating private rooms or information limited to certain customers.

In the next chapter, we will deal with the different techniques for creating rooms and their optimal management. All the knowledge acquired about a database will help us to create a simple chat that will allow us to maintain private conversations between two clients, restrict groups, or broadcast to all those connected.

5

Separating Communication in Rooms

Channels allow us to broadcast, send, and receive asynchronous messages to/from all clients that belong to a group. Within a group, we cannot filter a selection of users. To solve this problem and create a division or categories, we must resort to creating new Channels and manually grouping the clients.

So far, we can communicate with a customer who is isolated in a Channel or with all the customers connected to a common Channel. Now, it is time to learn how to control groups/Channels to separate and move customers between different groups as needed. You can even assign the same customer to several groups at the same time. For example, if we are creating a Chat, it would be useful for the user to be subscribed to a unique Channel to receive notifications, as well as another public group where all the customers can write freely, and other private groups where they can have conversations with other users. It makes sense for a client to receive or send different messages from various groups for different purposes.

In this chapter, we will cover the following topics:

- Basic functions for managing Channels
- Creating a full Chat

Technical requirements

You can download the code for this chapter from this book's GitHub repository: `https://github.com/PacktPublishing/Building-SPAs-with-Django-and-HTML-Over-the-Wire/tree/main/chapter-5`.

We will be using the template we constructed in *Chapter 4, Working with the Database*: `https://github.com/PacktPublishing/Building-SPAs-with-Django-and-HTML-Over-the-Wire/tree/main/chapter-4/initial-template`

I have changed the name of the application to Chat. Ensure that the App folder is called /app/chat/ and that apps.py has been renamed with its name variable:

```
from django.apps import AppConfig

class SimpleAppConfig(AppConfig):
    default_auto_field = "django.db.models.BigAutoField"
    name = "app.chat" # Update
```

If you rename an application, you must reflect this in /project_template/settings.py:

```
INSTALLED_APPS = [
    "channels",
    "django.contrib.admin",
    "django.contrib.auth",
    "django.contrib.contenttypes",
    "django.contrib.sessions",
    "django.contrib.messages",
    "django.contrib.staticfiles",
    "app.chat", # Update
]
```

I have also changed the name of the Consumer to ChatConsumer:

```
# app/chat/consumers.py
class ChatConsumer(JsonWebsocketConsumer):
```

In project_template/urls.py, you must change the View import:

```
from app.chat import views
```

In Caddyfile, change the domain from hello.localhost to chat.localhost:

```
http://chat.localhost {

    root * /usr/src/app/
```

Finally, remember that whenever you change the name of the Consumer, you must modify
`/project_template/asgi.py`:

```
from app.chat.consumers import ChatConsumer # Update
...
                    re_path(r"^ws/chat/$", ChatConsumer.
                        as_asgi()), # Update
...
```

With the template in place, we can now start the project, which will involve creating a Chat tool.

We will prepare the database with a model and generate some random users. However, before we continue, we must know about the functions that Channels provide for sending information to customers or for managing groups.

Basic functions for managing Channels

The basic functions for managing Channels are as follows:

- `send()`: This is used to send new messages from the Consumer to a single client. We have used this function from the beginning of this book. However, we used the `JsonWebsocketConsumer` wrapper to make `send_json()` more convenient for sending JSON:

  ```
  data = {
          "my_data": "hi",
          }
  self.send_json(data)
  ```

- `group_send()`: This is used to send new messages from the Consumer to a group of clients that we have previously defined. It is an asynchronous function, so we will need the whole Consumer to be asynchronous or, preferably, use the `async_to_sync` function. In the following example, you can see how the `{ "my_data": "hi" }` JSON is sent to the whole group as `"Main"`:

  ```
  from asgiref.sync import async_to_sync

  async_to_sync(self.channel_layer.group_send)(
          "Main", {
              "type": "send.hi", # Run "send_hi()"
                  method
              "my_data": "hi",
          }
  ```

```
def send_hi(self, event):
        """Event: Send "hi" to client"""
        data = {
            "my_data": event["my_data"],
        }
        self.send_json(data)
```

- `group_add()`: This is used to add a client to a new or existing group. The function is also asynchronous, so we will use `async_to_sync` again. In the following example, we are adding (`self.channel_name`) to a group called "Main":

```
async_to_sync(self.channel_layer.group_add)("Main", self.
channel_name)
```

- `group_discard()`: This is used to remove a client from a group. Again, this is an asynchronous function, so we are forced to use `async_to_sync`. In this example, we have removed (`self.channel_name`) from a group called "Main":

```
async_to_sync(self.channel_layer.group_discard)("Main",
self.channel_name)
```

With these functions, we can now dominate the world, or at least the world of real time. They are ideal for building a complete chat. And... why don't we compete with WhatsApp or Slack? They have hundreds of the best engineers, but we'll use Django here: it's a balanced fight. We will create a piece of software that will use the full potential of Channels to manage different groups with the following features:

- Group and public messages with no limit regarding the number of clients

- Private messages that can be sent between two clients

- Control over connected or disconnected clients

- Registered users can be identified

If we add the functions of Channels to those of Django, we will see that we have everything we need to manage the information and connect to the database. However, we need to learn about a few important details before connecting to Django's models. How can we isolate users?

Creating a full Chat

A very popular exercise when implementing WebSockets in any technology is to create a simple Chat. However, the difficulty increases considerably when we have several connected clients who are going to talk in private spaces and open groups so that any client can read or participate. Using Channels, we are creating a solid enough abstraction so that we can focus on other issues.

Let's create a Chat complete with modern features:

- Message history
- Private conversations
- Groups
- Customers associated with a registered user in the database

Next, we must define the database. We will define the models for users, rooms, and messages. That way, we will be able to store the actions of each user and there will be a record of everything that happens.

Defining the database

In this section, we are going to create some models in the database to manage customers, groups (which we will call rooms), and messages.

Edit /app/chat/models.py with the following content:

```python
from django.db import models
from django.contrib.auth.models import User


class Client(models.Model):
    """
    Clients for users
    """
    user = models.ForeignKey(User, on_delete=models.CASCADE)
    channel = models.CharField(max_length=200, blank=True,
null=True, default=None)
    created_at = models.DateTimeField(auto_now_add=True)

    def __str__(self):
        return self.user.username
```

The `Client` model allows us to have a record of the users that are connected or disconnected. It also allows us to store the private Channel of each client in case we need to send them an individual message from anywhere in the code:

```python
class Room(models.Model):
    """
    Rooms for users
    """
    users_subscribed = models.ManyToManyField(User,
        related_name="users_subscribed")
    clients_active = models.ManyToManyField(Client,
        related_name="clients_active")
    name = models.CharField(max_length=255, blank=True,
        null=True, default=None)
    is_group = models.BooleanField(default=False)

    def __str__(self):
        return self.name
```

Rooms will be a record of all the Channels that have been created and the clients that are subscribed to them via the `users_subscribed` column. We must perform this functionality because Channels do not allow us to access this information unless we use a third-party extension or make a record in the database, which is exactly what we are doing here. We'll use `clients_active` to know which clients are currently viewing the group, as they may be added, but at the same time disconnected or present in another room. This way, we will only send the updates or new HTML that's generated with the list of messages to the active clients and not to all the subscribed ones. Finally, `name` will be the name of the group and `is_group` will mark if this is a public group with many clients (`True`) or a private room (`False`), which is mandatory for controlling unwanted guests:

```python
class Message(models.Model):
    """
    Messages for users
    """
    User = models.ForeignKey(User, on_delete=models.
        CASCADE)
    room = models.ForeignKey(Room, on_delete=models.
        CASCADE)
```

```
text = models.TextField()
created_at = models.DateTimeField(auto_now_add=True)

def __str__(self):
    return self.text
```

The `Message` model will be in charge of storing the Chat messages. Each element will have an author (who we will call `user`), a Channel where a message has been sent (which we will call `room`), and the text of the message itself (which we will call `text`). In addition, we have added `created_at` to sort the messages when listing them.

We will lift `docker-compose` to carry out migrations:

docker-compose up

With the models defined, we are going to create the migration. We need to go into the `django` container and look for its name. As a hint, we know that it will end with `_django_1`:

docker ps

You will see a list of all your active containers, along with the processes they are running:

```
→ ~ docker ps
CONTAINER ID    IMAGE                     COMMAND
                NAMES
7ebde735cdf7    caddy:alpine              "caddy run --config …"
p, 2019/tcp     chapter-5_caddy_1
2c53f368f65c    chapter-5_django          "bash ./django-launc…"
                chapter-5_django_1
34eb8265f403    mailhog/mailhog:latest    "MailHog"
                chapter-5_mailhog_1
a8b7f362c4ba    redis:alpine              "docker-entrypoint.s…"
                chapter-5_redis_1
f228012b5226    postgres                  "docker-entrypoint.s…"
                chapter-5_postgresql_1
```

Figure 5.1 – Listing all the names of the containers after Docker is up

In my case, Django is `chapter-5_django_1`.

Now, let's enter the interactive Bash terminal:

```
docker exec -it chapter-5_django_1 bash
```

Here, we can create the necessary migrations:

```
./manage.py makemigrations chat
./manage.py migrate
```

With the database ready, we will include some random users to differentiate the clients.

Generating random users

Without registered users, we can't work, so let's create a Python script that makes some random data.

We will create a file called make_fake_users.py at the root of the project that contains the following content. At the moment, we will not be able to run it because we do not have **Faker** installed:

> **Faker**
>
> Faker is a Python library for generating fake data for various uses. Among its most common uses, it is used to insert data into the database to develop, prototype, or stress test an application.

```python
# make_fake_users.py
from django.contrib.auth.models import User
from faker import Faker


fake = Faker()

# Delete all users
User.objects.all().delete()


# Generate 30 random emails and iterate them.
for email in [fake.unique.email() for i in range(5)]:
    # Create user in database
    user = User.objects.create_user(fake.user_name(),
        email, "password")
    user.last_name = fake.last_name()
    user.is_active = True
    user.save()
```

Using Faker, we generate five unique emails. Then, we iterate them and create a unique user with a generated username and last name.

To install Faker, add the following line to the `requirements.txt` file:

```
# Fake data
Faker===8.13.2
```

Don't forget to recreate the Django image again so that the new dependency is installed from `Dockerfile`.

Now, let's run the Python script from Bash from the Django container:

```
./manage.py shell < make_fake_users.py
```

We currently have five random users ready to be used.

With the database created and populated with data, we can focus on generating the HTML and its components that will make use of this information.

Integrating HTML and styles

We need to display some nice minimalistic HTML to make the Chat usable, although we won't win the best web design of the year award.

Let's create `app/chat/templates/index.html` with the following content:

```
{# app/chat/templates/index.html #}
{% load static %}
<! doctype html>
<html lang="en">
<head>
    <meta charset="UTF-8">
    <meta name="viewport" content="width=device-width,
        user-scalable=no, initial-scale=1.0, maximum-
            scale=1.0, minimum-scale=1.0">
    <title>Chat</title>
    {# CSS #}
    <link rel="stylesheet" href="{% static 'css/main.css' %}">
    {# JavaScript #}
    <script defer src="{% static 'js/index.js' %}">
```

```
        </script>
    </head>
```

Let's link the future CSS and JavaScript files:

```
<body
        data-host="{{ request.get_host }}"
        data-scheme="{{ request.scheme }}"
    >
```

Now, let's communicate the path that JavaScript will use to connect to host and check whether the connection is secure with scheme:

```
    <h1 class="title">Chat</h1>
    {# Login user name #}
    <h2 class="subtitle">I'm <span id="logged-user">
    </span></h2>
    <div class="container chat">
        <aside id="aside">
            {# List of groups and users #}
            {% include "components/_aside.html" with
            users=users %}
        </aside>
        <main id="main">
            {# Chat: Group name, list of users and form to
            send new message #}
            {% include "components/_chat.html" %}
        </main>
    </div>
</body>
</html>
```

The preceding block is divided into three parts:

- `<span id="logged-user"></span>`: This is used to display the client's name
- `<aside id="aside"></aside>`: A component that will list the name of the groups and users that will be clickable to dynamically jump between Channels (or Rooms)

- `<main id="main"></main>`: Contains another component that will render all existing or new messages with the respective form to publish a new message

Now, let's create all the components. Let's start with `/app/chat/templates/components/_aside.html`:

```
<nav>
    {# Group links #}
    <h2>Groups</h2>
    <ul class="nav__ul">
        <li class="nav__li">
            <a
                class="nav__link"
                href="#"
                data-group-name="hi"
                data-group-public="true"
            >
                #hi
            </a>
        </li>
        <li class="nav__li">
            <a
                class="nav__link"
                href="#"
                data-group-name="python"
                data-group-public="true"
            >
                #python
            </a>
        </li>
                <li class="nav__li">
            <a
                class="nav__link"
                href="#"
                data-group-name="events"
                data-group-public="true"
            >
```

```
                        #events
                    </a>
            </li>
            </li>
            <li class="nav__li">
                <a
                    class="nav__link"
                    href="#"
                    data-group-name="off-topic"
                    data-group-public="true"
                >
                    #off-topic
                </a>
            </li>
        </ul>
        {# End Group links #}
        {# Users links #}
        <h2> Users</h2>
        <ul class="nav__ul">
        {% for user in users %}
            <li class="nav__li">
                <a
                    class="nav__link"
                    href="#"
                    data-group-name="{{ user.username }}"
                    data-is-group="false"
                >
                    {{ user.username }}
                </a>
            </li>
        {% endfor %}
        </ul>
        {# End Users links #}
</nav>
```

To simplify this code, we have manually typed in the names of all the groups where several clients will be able to speak at the same time. You are free to generate them from the model.

Now, let's create `/app/chat/templates/components/_chat.html`:

```
<section class="messages">

    {# Name of the connected group #}
    <h2 id="group-name">{{ name }}</h2>

    {# List of messages #}
    <div class="messages__list" id="messages-list"></div>

    {# Form to add a new message #}
    <form action="" class="messages__new-message">
        <input type="text" class="input" name="message"
            id="message-text" />
        <input type="submit" id="send" class="button"
            value="Send" />
    </form>

</section>
```

The preceding code contains the three essential parts of any self-respecting chat room:

- The name of the group or Channel present at the time
- A list of messages
- A form for adding new messages

However, the list of messages is empty. Where is the loop with the HTML template? To be tidy, we have placed it in another component located in `app/chat/templates/components/_list_messages.html`, which contains the following code:

```
{% for message in messages %}

    {# Item message #}
    <article class="message__item">
        <header class="massage__header">
```

```
            {# Username #}
            <h3 class="message__title">{{
                message.user.username }}</h3>

            {# Date of creation #}
            <time class="message__time">{{
                message.created_at|date: "d/m/Y H:i"
                    }}</time>

        </header>
        <div>

            {# Text #}
            {{ message.text }}

        </div>
    </article>
    {# End Item message #}

{% endfor %}
```

Now that we have defined all the HTML for the chat, we just need to add some minimal styles to give it structure.

Defining CSS styles

In this section, we will create a style file in `static/css/main.css` with a few fixes to make the future Chat more comfortable to use:

```css
/* Global styles */
:root {
    --color__background: #f6f4f3;
    --color__gray: #ccc;
    --color__black: #000;
}

* {
```

```
        font-family: "Helvetica Neue", Helvetica, Arial, sans-
            serif;
        box-sizing: border-box;
    }


body {
    margin: 0;
    background-color: var(--color__background);
}
```

We will prepare some colors, provide a nice typeface (if you only take one thing from the book, always use Helvetica), and arrange body:

```
/* General classes for small components */

.container {
    margin: 0 auto;
    padding: 1rem 0;
    max-width: 40rem;
}

.button {
    display: inline-block;
    padding: 0.5rem 1rem;
    background-color: var(--color__gray);
    border: 0;
    cursor: pointer;
    text-decoration: none;
}

.button:hover {
    filter: brightness(90%);
}

.input {
```

```
    display: block;
    width: 100%;
    outline: none;
    padding: .5rem;
    resize: none;
    border: 1px solid var(--color__gray);
    box-sizing: border-box;
}
```

We will slightly modernize the inputs and prepare a container to center the Chat:

```
/* Styles for chat */

.title {
    text-align: center;
}

.subtitle {
    text-align: center;
    font-weight: normal;
    margin: 0;
}

.chat {
    display: grid;
    grid-template-columns: 1fr 3fr;
    gap: 1rem;
}
```

Now, let's align <aside> and <main> horizontally:

```
/* Aside */

.nav__ul {
    list-style: none;
```

```
    padding: 0;
}

.nav__link {
    display: block;
    padding: 0.5rem 1rem;
    background-color: var(--color__gray);
    border: 1px solid var(--color__background);
    color: var(--color__black);
    text-decoration: none;
}

.nav__link:hover {
    filter: brightness(90%);
}

/* End Aside */
```

Here, we have fixed the browser and the links included within `<aside>` so that they have a clickable area that is comfortable enough to click on:

```
/* Chat */

.messages {
    display: grid;
    height: 30rem;
    grid-template-rows: 4rem auto 2rem;
}

.massage__header {
    display: grid;
    grid-template-columns: 1fr 1fr;
    grid-gap: 1rem;
}
```

```css
.messages__list {
    overflow-y: auto;
}

.message__item {
    border: 1px solid var(--color__gray);
    padding: 1rem;
}

.massage__header . message__title {
    margin-top: 0;
}

.massage__header . message__time {
    text-align: right;
}

.messages__new-message {
    display: grid;
    grid-template-columns: 8fr 1fr;
}

/* End Chat */
```

Finally, we have converted each chat message into a well-delimited box with a border. We also horizontally aligned the input and the form button to display it as we are used to today.

Now, we must create a view to render all the pieces we have created – the database, generated users, template, and HTML components – with a little CSS.

Creating the view

There's nothing at the root of the chat yet. Without a view and a route, the template cannot be served to the client. Even if we show a static template, we must indicate the path where it can be visited and rendered. We need a view to generate its presentation HTML.

In /app/chat/views.py, we will create a view called index that renders index.html with all the users, which will be displayed in <aside>:

```
from django.shortcuts import render
from django.contrib.auth.models import User


def index(request):
    """View with chat layout"""
    return render(
        request, "index.html", { "users":
            User.objects.all(). order_by("username")})
```

In /project_template/urls.py, we will add the view to be displayed when a visitor enters the site's root:

```
from django.urls import path
from app.chat import views

urlpatterns = [
    path("", views.index, name="index"),
]
```

Now, we will open the browser we have at hand with the domain of the project. The address is described in the DOMAIN variable of the docker-compose.yaml file. If you haven't touched the document, the address will be http://hello.localhost. In my case, I have changed it to http://chat.localhost.

We will be able to see in the browser the list of groups, written manually, and the list of existing users. In addition, we have a form where we can write future messages:

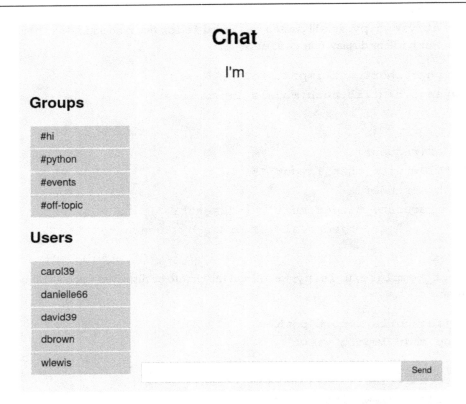

Figure 5.2 – The chat without any messages, group name, or client name

The visual part is ready; we can now focus all our attention on the chat logic. We already have the body; now, we need a brain to manage the logic.

Declaring JavaScript to listen to messages or send them

Let's update the `static/js/index.js` file with the following code:

```
    VARIABLES
*/
// Connect to WebSockets server (SocialNetworkConsumer)
const myWebSocket = new WebSocket(`${document.body.dataset.
scheme === 'http' ? 'ws' : 'wss'}://${ document.body.dataset.
host }/ws/chat/`);
```

We will connect to the backend with the WebSockets client using the scheme and host values we have printed in the dataset's <body> tag:

```
    FUNCTIONS
*/


/**
* Send data to WebSockets server
* @param {string} message
* @param {WebSocket} webSocket
* @return {void}
*/
function sendData(message, webSocket) {
    webSocket.send(JSON.stringify(message));
}
```

Let's retrieve the sendData() function we used in the previous examples to send messages to the backend:

```
/**
* Send message to WebSockets server
* @return {void}
*/
function sendNewMessage(event) {
    event.preventDefault();
    const messageText = document.querySelector('#message-
        text')
    sendData({
            action: 'New message',
            data: {
                message: messageText.value
            }
        }, myWebSocket);
    messageText.value = '';
}
```

Now, we must declare a function to send a new message. We won't need anything more than the text since I will know who the author is due to the Consumer:

```
/**
* Requests the Consumer to change the group with respect to the
Dataset group-name.
* @param event
*/
function changeGroup(event) {
    event.preventDefault();
    sendData({
            action: 'Change group',
            data: {
                groupName: event.target.dataset.groupName,
                isGroup: event.target.dataset.groupPublic
                    === "true".
            }
        }, myWebSocket);
}
```

The changeGroup() function will tell the Consumer to change the group and send us the HTML for the existing messages of the group. We will accompany this request with the dataset that stores the name of the Room to change and information about whether it is a multi-user group or a private conversation.

The final JavaScript fragment is for the backend listener:

```
    EVENTS
*/
// Event when a new message is received by WebSockets
myWebSocket.addEventListener("message", (event) => {
    // Parse the data received
    const data = JSON.parse(event.data);
    // Renders the HTML received from the Consumer
    document.querySelector(data.selector).innerHTML =
        data.html;
```

As in the previous examples, we will collect the JSON, parse it, and inject the HTML:

```
// Scrolls to the bottom of the chat
const messagesList = document.querySelector('#messages-
    list');
messagesList.scrollTop = messagesList.scrollHeight;
```

Every time we print a list of messages or receive a new message, the scroll will be placed at an inappropriate height. It may not scroll at all, or it may hang in the middle. To fix this, after each HTML injection, we must scroll down to the end of the element, always displaying the last message. This is a common behavior in all Chats:

```
/**
 * Reassigns the events of the newly rendered HTML
 */
// Button to send new message button
document.querySelector('#send').addEventListener('click',
sendNewMessage);
// Buttons for changing groups
document.querySelectorAll(".nav__link").forEach(button => {
    button.addEventListener("click", changeGroup);
});
});
```

Finally, we must reassign all the events after each render. The button that sends a new message, with an ID of send will execute sendNewMessage(), while all <aside> buttons will call changeGroup().

With the frontend defined, it's time to work with the Consumer. The Consumer is responsible for managing the database, listening to JavaScript, rendering the HTML, and managing the groups.

Building a Consumer to manage groups

In this section, we are going to define what will happen when a client connects, disconnects, sends us the action of changing groups, or adds a new message.

Edit `app/chat/consumers.py` with the following content:

```python
# app/chat/consumers.py
from channels.generic.websocket import JsonWebsocketConsumer
from django.template.loader import render_to_string
from asgiref.sync import async_to_sync
from channels.auth import login, logout
from django.contrib.auth.models import User
from .models import Client, Room, Message
```

Let's import the authentication system, User, and models:

```python
class ChatConsumer(JsonWebsocketConsumer):
```

The first thing we will do when we load the Consumer is delete Zombie Clients in case we forcefully close Django:

```python
    Client.objects.all().delete()

    def connect(self):
        """Event when client connects"""
```

Now, we will accept the customer's connection:

```python
        self.accept()
```

Next, we will obtain a random user who is not already registered as a customer:

```python
        user = User.objects.exclude(
            id__in=Client.objects.all().values("user")
        ).order_by("?").first()
```

Here, we will identify the user. It will be easier to work with sessions than storing the user ID:

```python
        async_to_sync(login)(self.scope, user)
        self.scope["session"].save()
```

Now, we will send the name of the registered user to the frontend:

```
self.send_html(
    {
        "selector": "#logged-user",
        "html": self.scope["user"].username,
    }
)
```

Next, we will register the client in the database to control who is connected:

```
Client.objects.create(user=user,
    channel=self.channel_name)
```

At this point, we will assign the "hi" group as the first room to be displayed when you enter. We have created a special function to handle some repetitive tasks when changing rooms. We will describe how the function works shortly:

```
self.add_client_to_room("hi", True)
```

Now, let's list the messages of the room where we have just assigned the client:

```
self.list_room_messages()
def disconnect(self, close_code):
    """Event when client disconnects"""
```

When a client disconnects, we will perform the following three tasks:

- Remove the client from the current room:

```
self.remove_client_from_current_room()
```

- Deregister the client:

```
Client.objects.get(channel=self.channel_name).delete()
```

- Log the user out:

```
logout(self.scope, self.scope["user"])
```

With that, we have automatically implemented a system that creates a session for the user, which is very handy for identifying and sending individual messages to the user, and also closes the user's session when the WebSocket client disconnects.

The function we have used in other examples for managing frontend actions is useful here. The backend tasks are simple: listen for and return JSON. We will always use the same functions, regardless of the application:

```python
def receive_json(self, data_received):
    """

        Event when data is received
        All information will arrive in 2 variables:
        "action", with the action to be taken
        "data" with the information
    """
    # Get the data
    data = data_received["data"]
```

Depending on the action, we will do one task or the other. These are the actions that are requested by the frontend, such as adding a new message or listing all messages.

We will only change groups if the frontend makes a request. But when will this request be made? When the user clicks on the name of the group where they want to go or on the user they want to talk to. The event will be captured by the frontend and the Change group action will be sent to the backend.

We can't work in the same way with a user who wants to enter a private room, where there will only be two users, and another user who will enter a public room (with no limit regarding users and open messages). The code is different. To control this situation, we will ask the frontend to send us isGroup. If it's true, it's a public group. If it's false, it is a private group between two users.

We will start by changing groups:

```python
match data_received["action"]:
    case "Change group":
        if data["isGroup"]:
```

If isGroup is True, we will add the client in a multi-user room: #hi, #python…

```python
            self.add_client_to_room(data["groupName"],
  data["isGroup"])
        else:
```

If isGroup is False, we will add a target user and the current user to the private room.

The major problem that we face is when two clients need to talk to each other, we need to ensure that we create a Room just for them. The difficulty is that we need to check if a Room already exists, and if it doesn't, we need to create a group and then inform the participants of this when they want to connect. We will have to make a decision tree, as follows:

1. Search for an already created Room where both clients have already spoken in the past. If it exists, retrieve the name of the Room and add the client to the group. If it does not exist, go to *Step 2*.

2. See if the users who want to talk to each other are alone in a Room. This is because they have created a Room and are waiting for another user to join and talk to them. If not, go to *Step 3*.

3. Create a new Room and hope that a user wants to talk to us.

First, we will search for rooms where both clients match:

```
            room = Room.objects.filter(users_
subscribed__in=[self.scope["user"]], is_group=False).
intersection(Room.objects.filter(users_subscribed__in=[user_
target], is_group=False)).first())
            if room and user_target and room.users_
subscribed.count() == 2:
```

Then, we will get the client who wants to talk:

```
            user_target = User.objects.
filter(username=data["groupName"]).first()
```

An existing group may be found where both the target and current clients are already talking. This is the most favorable case as there is a previous conversation where a Room has already been created. In this case, the client can be added to the group to talk:

```
            self.add_client_to_room(room.name)
        else:
```

If there is no existing group, look for a room where the target user is alone:

```
            room = Room.objects.filter(
                users_subscribed__in=[
                    user_target,
                ],
                is_group=False,
            ).last()
            if room and room.users_subscribed.
count() == 1:
```

If there is a room, let's join:

```
                    self.add_client_to_room(room.name)
            else:
```

If we have not found a room where the target user is alone, we must create a new room:

```
            self.add_client_to_room()
```

After moving the client to another room, we need to give them feedback so that they know which room they are in at the moment. We will send them the name of the room:

```
        self.send_room_name()
    case "New message":
```

Here, we have received a new message to save:

```
        self.save_message(data["message"])
```

After this, there will be changes to show to the client, such as adding a new message. We will always send a list of the messages where the client is located so that they have the latest changes in the HTML:

```
    self.list_room_messages()


def send_html(self, event):
    """Event: Send html to client"""
    data = {
        "selector": event["selector"],
        "html": event["html"],
    }
    self.send_json(data)


def list_room_messages(self):
    List all messages from a group""""""
    room_name = self.get_name_room_active()
    # Get the room
    room = Room.objects.get(name=room_name)
    # Get all messages from the room
```

```
        messages = Message.objects.filter(room=room). order_
by("created_at")
        # Render HTML and send to client
        async_to_sync(self.channel_layer.group_send)(
            room_name, {
                "type": "send.html", # Run "send_html()" method
                "selector": "#messages-list",
                "html": render_to_string("components/_list_
messages.html", {"messages": messages})
            }


    def send_room_name(self):
        """Send the room name to the client"""
        room_name = self.get_name_room_active()
        room = Room.objects.get(name=room_name)
        data = {
            "selector": "#group-name",
            # Concadena # if it is a group for aesthetic
reasons
            "html": ("#" if room.is_group else "") + room_name,
        }
        self.send_json(data)
```

Whenever we want to know who we are, we can use `self.scope["user"]`. It will return the logged-in user object:

```
    def save_message(self, text):
        "Save a message in the database""""""
        # Get the room
        room = Room.objects.get(name=self.get_name_room_
active())
        # Save message
        Message.objects.create(
            user=self.scope["user"],
            room=room,
```

```
            text=text,
```

To add a user to a Room, we must do the following:

1. Get the user client.

2. Remove the client from the previous room.

3. Get or create a room.

4. If the room has no name, it is assigned `"private_{id}"`. For example, if `id` is 1, it shall be `"private_1"`.

5. Add a client to the group.

6. Send the group name to the client, as follows:

```python
    def add_client_to_room(self, room_name=None, is_
group=False):
        """Add customer to a room within Channels and save the
reference in the Room model."""""
        client = Client.objects.get(user=self.scope["user"])
        self.remove_client_from_current_room()
        room, created = Room.objects.get_or_create(name=room_
name, is_group=is_group)
        if not room.name:
            room.name = f "private_{room.id}"
            room.save()
        room.clients_active.add(client)
        room.users_subscribed.add(client.user)
        room.save()
        async_to_sync(self.channel_layer.group_add)(room.name,
self.channel_name)
        self.send_room_name()
```

Let's describe the preceding code in more detail. There are several important parts to understand:

1. Obtaining the name of the Room where we are active is relatively easy by filtering the database:

```python
def get_name_room_active(self):
    """Get the name of the group from login user"""
    room = Room.objects.filter(clients_active__user_
id=self.scope["user"].id).first()
    return room.name
```

2. To remove ourselves from a group, we must do the reverse:

```python
def remove_client_from_current_room(self):
    Remove client from current group""""""
```

3. We get all the Rooms where we are active:

```python
    client = Client.objects.get(user=self.
scope["user"])
    rooms = Room.objects.filter(clients_active__
in=[client])
```

4. We go in and out of the Room and eliminate each other:

```python
    for room in rooms:
```

5. We remove the client from the group:

```python
        async_to_sync(self.channel_layer.group_
discard)(room.name, self.channel_name)
```

6. We remove the client from the Room model:

```python
        room.clients_active.remove(client)
        room.save()
```

And the Chat is now complete. When we enter, it will render our username and the name of the Room where we are active. At the start of the chat, we will see #hi:

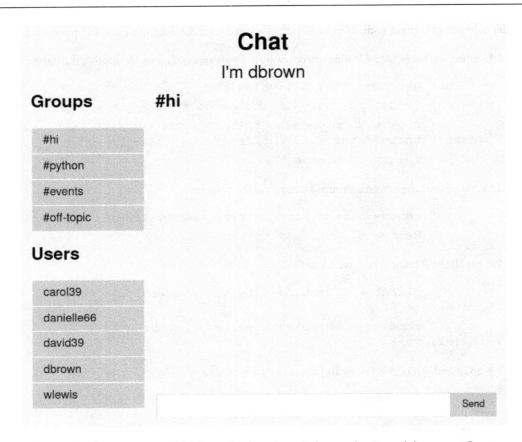

Figure 5.3 – Our username, which is randomly assigned when we log in, and the current Room

If we open a tab in another browser or use private browsing, a new random user will be assigned to the session, and we will be able to post to any of the groups. All the messages will be rendered in real time to the clients that are present or active in the group:

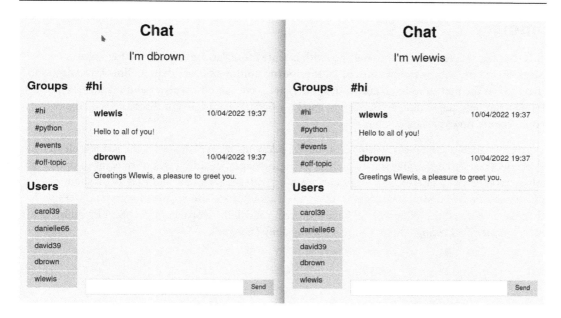

Figure 5.4 – Any user can write freely in the groups, without limitations

If we open a third browser, we can experience private rooms or conversations between two clients:

Figure 5.5 – A private conversation between two users

At any time, we can exchange messages with existing users or groups. Moreover, as we have a database to store the messages, even if we restart the Docker containers, we will always see the history with everything written, sorted by creation date. Here, we have a realistic Chat, with real-time response and logic in the backend. It's amazing what we can achieve if we know how to use Django's native tools and know how to manage Channels.

Summary

In this chapter, we created a functional chat with private rooms and groups, like other software such as Slack or Teams, with very few lines of JavaScript (no comments, less than 35 lines). In addition, we have taken the first steps in an authentication system. We can now register and manage clients in different Channels, depending on our needs, and know who is connected or disconnected. The magic is over – we are now masters of Channels.

In the next chapter, *Chapter 6, Creating SPAs on the Backends*, we will deal with the last few elements that are necessary to dynamize a site, such as changing pages, deciding when we want to update a whole section or add a new HTML fragment, working with sessions so as not to depend so much on the database, and validating the origin of the data to avoid **cross-site request forgery** (**CSRF**) with WebSockets. With all the skills we will have acquired, we will develop a complete SPA by building a blog in *Chapter 7, Creating a Real-Time Blog Using Only Django*.

Part 3:
HTML over
WebSockets

In this part, we will assimilate a modern architecture for creating SPAs using the backend to render the HTML and send it via a WebSockets connection to the frontend. We will use everything we have learned to create a real-time SPA Blog where all the load and logic will be in the backend, leaving the frontend with the only responsibility of handling events.

In this part, we will cover the following chapters:

- *Chapter 6, Creating SPAs on the Backends*
- *Chapter 7, Creating a Real-Time Blog Using only Django*

6
Creating SPAs on the Backends

We cannot create a complete site by simply managing groups and sending HTML to the client. We must first master a variety of small solutions in order to be able to build a dynamic page that interacts with the user, with essential features such as page switching!

When the first **single-page applications** (**SPAs**) were created, the developers at the time were forced to spend many hours on functionalities that had been free when using the HTTP protocol: routing, sessions, authentication, or origin verification, among others. Poor them! They had to re-invent the wheel with a rebellious adolescent JavaScript that was not very cross-browser compatible. However, they survived, or so I would like to think, by defining techniques in the frontend that have managed to mimic the same behavior as HTTP; these techniques have lasted until today. For example, in a routing system, when a SPA redraws a screen, the browser URL is modified to put the user in context. On the other hand, if a visitor manually types the address, the application reacts by loading the components that make up the screen. All tasks are exhausting to implement with JavaScript. It is not cheap to change content without making a new request. If we just used plain HTML, we wouldn't need to do anything, but of course, the user would experience a page refresh with every change. And what does all this have to do with us? If we create pages using the WebSockets protocol, we find ourselves in a similar situation; we have to invent formulations to simulate the behavior that a polite user expects from us.

Channels is simple in functionality compared to other libraries, but at the same time very mature and consistent with the real world. It is a framework born out of necessity. It relies on Django to give us the essentials to solve typical problems while providing flexibility.

In this chapter, we are going to review different approaches to the following:

- Switching between pages
- Server-side rendering for each route
- Including a browser to achieve dynamic navigation

- Changing URLs
- Hydrating sections or components
- Creating temporary sessions for sessions
- Avoiding **cross-site request forgery (CSRF)** with WebSockets

Therefore, we are going to focus on how to solve each point in order to prepare ourselves for *Chapter 7, Creating a Real-Time Blog Using Only Django*.

Let's organize our projects a bit better. From now on, we are going to divide Channels into two files: `consumers.py`, which will be the equivalent of `views.py` or a communication gateway between the frontend and the backend, and `actions.py`, where the logic or functions will be located.

We'll start by adding a complete system for switching pages. You will not need to follow each point in order, as you will find examples of how we can solve each task, not a tutorial.

Technical requirements

All the code from the different sections can be found at the following link:

`https://github.com/PacktPublishing/Building-SPAs-with-Django-and-HTML-Over-the-Wire/tree/main/chapter-6`

Switching between pages

At some point, the user will need to go to another page or change context. We are going to make them think this is happening, but in reality, it is going to be a magic trick since really, they will never move from the first HTML we gave them at the beginning. However, and here's the key, they will perceive that the page is being changed. To achieve this deception (sorry, achievement), we will carry out the following tasks:

1. Change the HTML of the main content or everything that belongs to `<main>`. Meanwhile, we will always keep the static sections of the pages, such as `<header>`, `<aside>`, or `<footer>`.

2. Implement server-side rendering to render the HTML belonging to each URL.

3. Visually mark in the `<nav>` where we are with a CSS style.

4. Modify the browser URL via the JavaScript API. It is an aesthetic change but the URL acts as breadcrumbs to guide the visitor.

The objective is to build a site with three pages: **Home**, **Login**, and **Signup**. We'll start with an HTML layout that we will call `base.html`:

```
{% load static %}
<!doctype html>
```

```
<html lang="en">
<head>
    <meta charset="UTF-8">
    <meta name="viewport" content="width=device-width,
        user-scalable=no, initial-scale=1.0, maximum-
            scale=1.0, minimum-scale=1.0">
    <title>Example website</title>
    <link rel="stylesheet" href="{% static 'css/main.css'
        %}">
    <script defer src="{% static 'js/index.js' %}">
    </script>
</head>
<body
        data-host="{{ request.get_host }}"
        data-scheme="{{ request.scheme }}">
    <div class="container">
        <header>
            <nav id="nav" class="nav">{% include
                'components/_nav.html' %}</nav>
        </header>
        <main id="main">{% include page %}</main>
        <footer class="footer">My footer</footer>
    </div>
</body>
</html>
```

The components/_nav.html component will be discussed later when we talk about navigation. The important thing is that we are going to incorporate an include inside <main> that we will use to create a future server-side rendering system.

Next, in the Consumer class, we will create the "Change page" action, which will invoke the send_page (self, "page name") function inside actions.py:

```
# app/app_template/consumers.py
from channels.generic.websocket import JsonWebsocketConsumer
import app.app_template.actions as actions
```

```python
class ExampleConsumer(JsonWebsocketConsumer):

    def connect(self):
        """Event when client connects"""
        # Accept the connection
        self.accept()

    def disconnect(self, close_code):
        """Event when client disconnects"""
        pass

    def receive_json(self, data_received):
        """
            Event when data is received
            All information will arrive in 2 variables:
            "action", with the action to be taken
            "data" with the information
        """
        # Get the data
        data = data_received["data"]
        # Depending on the action we will do one task or
another.
        match data_received["action"]:
            case "Change page":
                actions.send_page(self, data["page"])

    def send_html(self, event):
        """Event: Send html to client"""
        data = {
            "selector": event["selector"],
            "html": event["html"],
            "append": "append" in event and event["append"],
            "url": event["url"] if "url" in event else "",
        }
```

```
        self.send_json(data)
```

As you may have noticed, `send_html` has also already been modified to incorporate `append`, which we will use to indicate whether we want to add a block of HTML to the selector or replace all the content (for the moment, we will not implement it), while `url` will be used to indicate the URL that will be displayed in the browser.

In `app/app_template/actions.py`, we would define the function that renders HTML and sends it to the frontend:

```python
from .forms import LoginForm, SignupForm
from asgiref.sync import async_to_sync
from django.template.loader import render_to_string
from django.urls import reverse
from datetime import datetime


def send_page(self, page):
    """Render HTML and send page to client"""

    # Prepare context data for page
    context = {}
    match page:
        case "login":
            context = {"form": LoginForm()}
        case "signup":
            context = {"form": SignupForm()}

    context.update({"active_nav": page})
```

We prepare the variables that will be used to render the HTML templates, the `Form` object corresponding to each page, and the name of the page where we are:

```python
    # Render HTML nav and send to client
    self.send_html({
        "selector": "#nav",
        "html": render_to_string("components/_nav.html",
            context),
    })
```

At each page change, we must redraw the `main` browser to mark where we are:

```
# Render HTML page and send to client
self.send_html({
    "selector": "#main",
    "html": render_to_string(f"pages/{page}.html",
        context),
    "url": reverse(page),
})
```

Finally, we send the HTML of the page to the frontend at `<main>` with a variable called `url`. This will be used by JavaScript later on to modify the address of the browser.

Before we continue to incorporate page switching, let's make a pitstop to incorporate the rendering of each view using Django. It will simplify the task of creating a browser that we'll need to move between pages.

Server-side rendering for each route

After preparing the `Consumer` class to change pages dynamically, we are going to incorporate a trivial system with Django for the management of routes and the rendering of each page without depending on Channels, so that crawlers can index the content. We'll define three templates (`home.html`, `login.html`, and `signup.html`).

The content of `app/app_template/templates/pages/home.html` will be a few lines of HTML:

```
<section>
    <h1>Welcome to an example of browsing with WebSockets over
the Wire.</h1>
    <p>You will be able to experience a simple structure. </p>
</section>
```

Then, on the second page, representing a login form, we will use a `form` object to list all the fields and then validate. This will be an argument that we will pass when rendering the template.

We write the following code in `app/app_template/templates/pages/login.html`:

```
<h1>Login</h1>
<form id="login-form">
    {{ form.as_p }}
```

```
            <input type="submit" class="button" value="Login">
</form>
```

Finally, we repeat the same structure in `app/app_template/templates/pages/signup.html` using a `form` object:

```
<h1>Signup</h1>
<form id="signup-form">
    {{ form.as_p }}
    <input type="submit" class="button" value="Signup">
</form>
```

Before defining the views, we need to structure the forms. In `app/app_template/forms.py`, we add the following content:

```python
from django import forms

class LoginForm(forms.Form):
    email = forms.CharField(
        label="Email",
        max_length=255,
        widget=forms.EmailInput(attrs={"id": "login-email",
            "class": "input"}),
    )
    password = forms.CharField(
        label="Password",
        max_length=255,
        widget=forms.PasswordInput(attrs={"id": "login-
            password", "class": "input"}),
    )


class SignupForm(forms.Form):
    username = forms.CharField(
        label="Username",
        max_length=255,
        widget=forms.TextInput(attrs={"id": "signup-
            username", "class": "input"}),
    )
```

```
email = forms.EmailField(
    label="Email",
    max_length=255,
    widget=forms.EmailInput(attrs={"id": "signup-
        email", "class": "input"}),
)
password = forms.CharField(
    label="Password",
    max_length=255,
    widget=forms.PasswordInput(attrs={"id": "signup-
        password", "class": "input"}),
)
password_confirm = forms.CharField(
    label="Confirm Password",
    max_length=255,
    widget=forms.PasswordInput(
        attrs={"id": "signup-password-confirm",
            "class": "input"}
    ),
)
```

With the templates and forms ready to be rendered, we edit app/app_template/views.py:

```
from django.shortcuts import render
from .forms import LoginForm, SignupForm


def home(request):
    return render(
        request,
        "base.html",
        {
            "page": "pages/home.html",
            "active_nav": "home",
        },
    )
```

```
def login(request):
    return render(
        request,
        "base.html",
        { "page": "pages/login.html", "active_nav":
            "login", "form": LoginForm()},
                )



def signup(request):
    return render(
        request,
        "base.html",
        { "page": "pages/signup.html", "active_nav": "signup",
"form": SignupForm()},
    )
```

In all cases, we use base.html as the main layout, where we will alter the content of <main> with the page variable:

```
<main id="main">{% include page %}</main>
```

The active_nav variable is a visual resource to inform the visitor via CSS where they are by changing the color of the appropriate hyperlink. We can ignore it for the moment.

Now, we edit project_template/urls.py to define all the paths:

```
from django.contrib import admin
from django.urls import path
from app.app_template import views

urlpatterns = [
    path("", views.home, name="home"),
    path("login/", views.login, name="login"),
    path("signup/", views.signup, name="signup"),
    path("admin/", admin.site.urls),
```

Nothing out of the ordinary; it's Django's own routing system. Except for one detail: we haven't extended it at any point. The usual way would be to render `home.html` and not `base.html`. In other words, `home.html` is the content of the page, which uses `base.html` as its structure:

```
{% extends 'base.html' %}
<section>
    <h1>Welcome to an example of browsing with WebSockets over
the Wire</h1>.
    <p>You will be able to experience a simple structure.</p>
</section>
```

We've done it this way because Django must adapt to the way we're going to draw HTML via WebSockets. We are only interested in modifying `<main>` and the templates must be isolated in components to work this way.

You can now open the three paths to see how they render without using a `Consumer` class.

We can see how the root of the site is rendered:

Figure 6.1 – Rendering the Home page using Django

The login form is rendered without any problem:

Figure 6.2 – Rendering the Login page using Django

The same happens when we render the registration page:

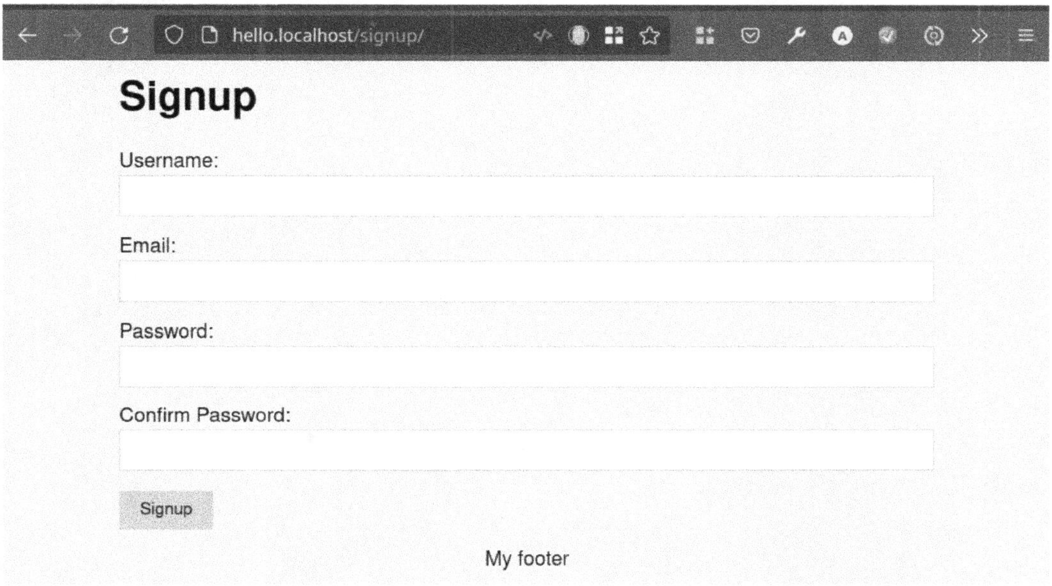

Figure 6.3 – Rendering the Signup page using Django

With the server-side rendering system in place, we are going to incorporate a browser that executes actions to dynamically modify the body of the page or small sections of it.

Including a browser to achieve dynamic navigation

After incorporating the templates, views, and routes for traditional navigation, we will create a dynamic navigation system.

We declare a file in the `app/app_template/components/_nav.html` path with the following content:

```
<ul class="nav__ul">
    <li>
        <a
            href="#"
            class="nav__link nav__link nav__link--
                page{% if active_nav == "home" %}
                    active{% endif %}"
```

```
                    data-target="home"
        >
            Home
        </a>
    </li>
    <li>
        <a
            href="#"
            class="nav__link nav__link--page{% if
            active_nav == "login" %} active{% endif %}"
            data-target="login"
        >
            Login
        </a>
    </li>
    <li>
        <a
            href="#"
            class="nav__link nav__link nav__link--
                page{% if active_nav == "signup" %}
                    active{% endif %}"
            data-target="signup"
        >
            Signup
        </a>
    </li>
</ul>
```

We will pass active_nav to the template with the name of the page we want to mark with CSS, adding the active class. On the other hand, data-target is a dataset that will collect JavaScript to send to the Consumer class and tell it which page to render.

In JavaScript, we will assign a click event to each <a> to send the action to change the desired page to the Consumer class. Which page? The one we have saved in the data-target. We must be careful before adding a new event listener; it is highly recommended that we delete the previous one to avoid duplicating the events to the same functions. Remember that the HTML is swapped but JavaScript remains static.

Edit `static/js/index.js`, adding the browser events:

```
/**
 * Send message to update page
 * @param {Event} event
 * @return {void}
 */
function handleClickNavigation(event) {
    event.preventDefault();
    sendData({
        action: 'Change page',
        data: {
            page: event.target.dataset.target
        }
    }, myWebSocket);
}


/**
 * Send message to WebSockets server to change the page
 * @param {WebSocket} webSocket
 * @return {void}
 */
function setEventsNavigation(webSocket) {
    // Navigation
    document.querySelectorAll('.nav__link--
        page').forEach(link => {
        link.removeEventListener('click',
            handleClickNavigation, false);
        link.addEventListener('click',
            handleClickNavigation, false);
    });
}
// Event when a new message is received by WebSockets
myWebSocket.addEventListener("message", (event) => {
    // Parse the data received
    const data = JSON.parse(event.data);
```

```
    // Renders the HTML received from the Consumer
    const selector = document.querySelector(data.selector);
    selector.innerHTML = data.html;
    /**
     * Reassigns the events of the newly rendered HTML
     */
    updateEvents();
});


/**
 * Update events in every page
 * return {void}
 */
function updateEvents() {
    // Nav
    setEventsNavigation(myWebSocket);
}



    INITIALIZATION
 */
updateEvents();
```

Now, we just need to add some CSS in `static/css/main.css` to change the color of the link for where we are:

```css
.nav__link.active {
    color: var(--color__active);
    text-decoration: none;
}
```

We can now switch between pages, although this is not reflected in the browser's address bar.

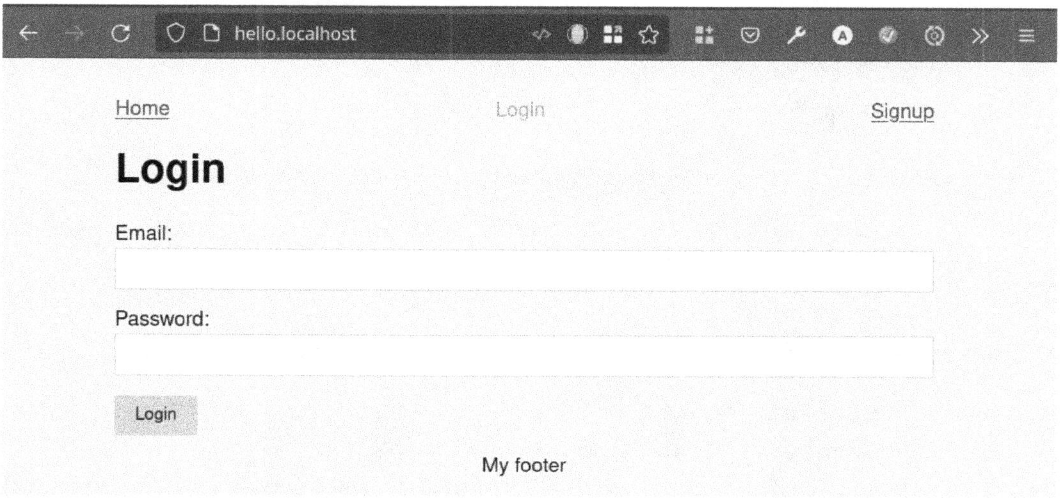

Figure 6.4 – Login page loaded with ability to navigate between Home and Signup dynamically

We have built a website with the ability to navigate between pages, as well as integrating traditional rendering to feed content to **search engine spiders**. However, we do not give feedback to visitors. The next objective will be to display the hierarchy and/or name of the page in the URL.

Changing URLs

We have managed to change pages and visually mark in the browser where we are, but the browser URL is still passive. We are going to add a mechanism to update the path every time we change pages.

In JavaScript, we can use the History API to manipulate the address that the visitor sees in the browser. For example, if you wanted to show that you are at `/login/`, you would implement the following:

```
history.pushState({}, '', '/login/')
```

What we will do is modify the event listener message by adding the line we just mentioned, together with a new parameter that will always send a `Consumer` class called `url`:

```
// Event when a new message is received by WebSockets
myWebSocket.addEventListener("message", (event) => {
    // Parse the data received
    const data = JSON.parse(event.data);
    // Renders the HTML received from the Consumer
    const selector = document.querySelector(data.selector);
```

```
    selector.innerHTML = data.html;
    // Update URL
    history.pushState({}, '', data.url) // New line
    /**
     * Reassigns the events of the newly rendered HTML
     */
    updateEvents();
});
```

In `Consumer`, we will modify the `send_html` function to support the `url` parameter:

```
def send_html(self, event):
        """Event: Send html to client"""
        data = {
            "selector": event["selector"],
            "html": event["html"],
            "url": event["url"] if "url" in event else "", #
New line
        }
        self.send_json(data)
```

While in `actions.py`, we'll modify `send_page` to send the route, but what is the route? Thanks to Django and `urls.py`, we can use `reverse`, which will return the full path:

```
from django.urls import reverse

def send_page(self, page):
...
        self.send_html({
        "selector": "#main",
        "html": render_to_string(f "pages/{page}. html",
            context),
        "url": reverse(page),
    })
...
```

We can now visualize the routes when navigating.

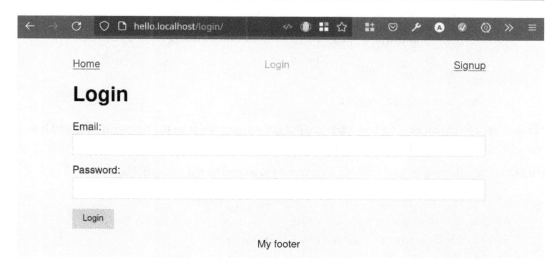

Figure 6.5 – Displaying the URL when browsing dynamically

We have a serious limitation though: we cannot add HTML blocks. We can only replace them. It is inefficient to render a whole page if we are only looking to add a new element to an existing list, for example. So, we are going to include a system that enables us to decide whether we are going to replace or add a piece of HTML to any available selector.

Hydrating sections or components

Although we have a function that can dynamically include HTML rendered from a template and apply it to a tag present in the document, we cannot decide whether we want to replace or insert HTML, in other words, hydrate or replace the DOM.

Hydration is a technique in web development where client-side JavaScript converts a static HTML web page into a dynamic web page by attaching event handlers to the HTML elements. This allows for a fast **First Contentful Paint** (**FCP**) but there is a period of time afterward where the page appears to be fully loaded and interactive. However, it is not until the client-side JavaScript is executed and event handlers have been attached.

To solve this problem, we will start by remembering that the Consumer class is prepared to receive the append instruction:

```
def send_html(self, event):
    """Event: Send html to client"""
    data = {
        "selector": event["selector"],
        "html": event["html"],
```

```
            "append": "append" in event and
                event["append"],
            "url": event["url"] if "url" in event else "",
        }
        self.send_json(data)
```

By default, append will be a `False` variable. But if the client sends us the append data and it is True, we will send what we want to add to the frontend and JavaScript will take care of the rest.

We include the following in `static/js/index.js`, a conditional to control append:

```javascript
myWebSocket.addEventListener("message", (event) => {
    // Parse the data received
    const data = JSON.parse(event.data);
    // Renders the HTML received from the Consumer
    const selector = document.querySelector(data.selector);
    // If append is received, it will be appended.
     Otherwise the entire DOM will be replaced.
    if (data.append) {
        selector.innerHTML += data.html;
    } else {
        selector.innerHTML = data.html;
    }
    // Update URL
    history.pushState({}, '', data.url)
    /**
     * Reassigns the events of the newly rendered HTML
     */
    updateEvents();
});
```

To check that it works, we are going to add a list of **Laps** to the **Home** page. A Lap is a unit of time that is stored inside a stopwatch as a history of recorded periods of time. For example, if it were a Formula 1 race, you could visualize how long each car took to complete a lap just by looking at the recorded lap time.

Each time a button is pressed, a new item with the current time will be added:

1. We edit the Home template hosted in app/app_template/templates/pages/home.
    html. We include a button and an unordered list:

    ```
    <section>
        <h2>Laps</h2>
        <p>
            <button id="add-lap">Add lap</button>
        </p>
        <ul id="laps"></ul>
    </section>
    ```

2. In JavaScript, hosted in the example in static/js/index.js, we incorporate the event
    into the button. It will just send an action without any data:

    ```
    /**
     * Send new Lap
     * @param {Event} event
     * @return {void}
     */
    function addLap(event) {
        sendData({
            action: 'Add lap',
            data: {}
        }, myWebSocket);
    }


    /**
     * Update events in every page
     * return {void}
     */
    function updateEvents() {
        // Nav
        setEventsNavigation(myWebSocket);
        // Add lap
        const addLapButton = document.querySelector('#add-
            lap');
    ```

```
        if (addLapButton !== null) {
            addLapButton.removeEventListener('click',
                addLap, false);
            addLapButton.addEventListener('click', addLap,
false);
        }
    }
```

3. In the Consumer class, in the app/app_template/consumers.py path, we capture the action and call a future add_lap function:

```
        def receive_json(self, data_received):
            """
                Event when data is received
                All information will arrive in 2 variables:
                "action", with the action to be taken
                "data" with the information
            """
            # Get the data
            data = data_received["data"]
            # Depending on the action we will do one task or
another.
            match data_received["action"]:
                case "Change page":
                    actions.send_page(self, data["page"])
                case "Add lap":
                    actions.add_lap(self)
```

4. In actions, located in app/app_template/actions.py, we include the function we called add_lap in the previous point. We include in the #laps selector the HTML fragment rendered in the _laps.html template, which is created from a variable called time, with the current time:

```
    def add_lap(self):
        """Add lap to Home page"""
        # Send current time to client
        self.send_html({
            "selector": "#laps",
            "html": render_to_string
```

```
            ("components/_lap.html",
                {"time": datetime.now()}),
        "append: True,
    })
```

5. Finally, we build the `app/app_template/templates/components/_lap.html` component:

```
<li>{{ time|date: "h:i:s" }}</li>
```

And that's it. We test how we can update the list over time by pressing the **Add lap** button in **Home**.

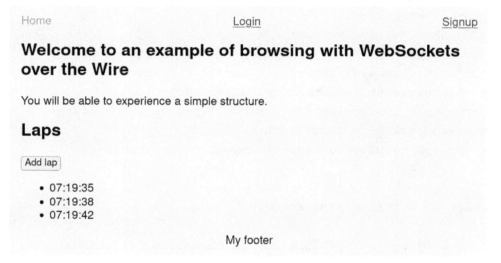

Figure 6.6 – Adding HTML snippets to preserve the previous content in an
unordered list that records the time when the button was clicked

We have improved the HTML rendering system to be more selective and efficient. We can now decide when we want to add or replace a DOM.

If you change the page and go back to **Home**, you will find that all the times have been deleted. To avoid this problem, we could save times in the database, or we can find an optimal solution by creating a temporary session for the user.

Creating temporary sessions for clients

To have unique sessions for each client, we will need to activate middleware that enables this feature. Channels provides us with `SessionMiddlewareStack` or `AuthMiddlewareStack`, which also include tools to build login or logout functionality. We will use `AuthMiddlewareStack` whenever we can.

We edit `project_template/asgi.py` as follows:

```python
import django

os.environ.setdefault("DJANGO_SETTINGS_MODULE", "project_
template.settings")
from django.conf import settings
django.setup()
from django.core.asgi import get_asgi_application
from channels.auth import AuthMiddlewareStack
from channels.routing import ProtocolTypeRouter, URLRouter
from django.urls import re_path
from app.app_template.consumers import ExampleConsumer


application = ProtocolTypeRouter(
    {
        # Django's ASGI application to handle traditional HTTP
requests
        "http": get_asgi_application(),
        # WebSocket handler
        "websocket": AuthMiddlewareStack(
            URLRouter(
                [
                    re_path(r"^ws/example/$", ExampleConsumer.
as_asgi()),
    }
```

We can now create sessions within the `Consumer` class with the following:

```python
self.scope["session"]["my name"] = "value".
self.scope["session"].save()
```

Getting it will be the same as reading from any Python dictionary:

```
print(self.scope["session"]["my name"])
# value
```

To exemplify its potential, we will create a classic to-do app on the home page. Even if we change pages, all the tasks that we left pending are always present when we come back, just like in real life. See the following:

1. At the end of the home template, we include an `<input>` to add the text of the task, a button to trigger the action, and the list where it will be displayed:

```
<section>
    <h2>TODO</h2>
    <input type="text" id="task">
    <button id="add-task"> Add task</button>
    <ul id="todo">{% include "components/_tasks.html"
with tasks=tasks %}</ul>
</section>
```

2. We will need a component that lists all the tasks. Therefore, in `app/app_template/templates/components/_tasks.html`, we include the following code:

```
{% for task in tasks %}
    {% include "components/_task-item.html" with
task=task %}
{% endfor %}
```

3. Inside the previous component, we use another component to render the item. We declare `app/app_template/templates/components/_task-item.html` with a `<li>` and the name of the task:

```
<li>{{ task }}</li>
```

4. In the `Consumer` class, when a user connects, we create a session called `tasks` with an empty list that we can fill in. On the other hand, we capture the action received from the frontend called "Add task" and call the `add_task` function in `actions.py`:

```
import app.app_template.actions as actions


class ExampleConsumer(JsonWebsocketConsumer):
```

```python
    def connect(self):
        """Event when client connects"""
        # Accept the connection
        self.accept()
        # Make session task list
        if "tasks" not in self.scope["session"]:
            self.scope["session"]["tasks"] = []
            self.scope["session"].save()

def receive_json(self, data_received):
        # Get the data
        data = data_received["data"]
        # Depending on the action we will do one task or
another.
        match data_received["action"]:
 # Other actions
            case "Add task":
                actions.add_task(self, data)
```

5. In `actions.py`, we declare the `add_task` function, which will add the task to the session, but we will also create `context` for home with the `session` variable:

```python
from .forms import LoginForm, SignupForm
from asgiref.sync import async_to_sync
from django.template.loader import render_to_string
from django.urls import reverse
from channels.auth import login, logout
from django.contrib.auth.models import User
from django.contrib.auth import authenticate
from datetime import datetime


def send_page(self, page):
    """Render HTML and send page to client"""

    # Prepare context data for page
```

```
        context = {}
        match page:
            case "home":
                context = {"tasks": self.scope["session"]
                    ["tasks"] if "tasks" in self.scope
                        ["session"] else []}
            case "login":
                context = {"form": LoginForm()}
            case "signup":
                context = {"form": SignupForm()}
...


def add_lap(self):
    "Add lap to Home page""""
    # Send current time to client
    self.send_html({
        "selector": "#laps",
        "html": render_to_string
            ("components/_lap.html", {"time":
                datetime.now()}),
        "append: True,
    })


def add_task(self, data):
    "Add task from TODO section""""
    # Update task list
    self.send_html({
        "selector": "#all",
        "html": render_to_string("components/_task-
            item.html", {"task": data["task"]}),
        "append: True,
    })
    # Add task to list
    self.scope["session"]["tasks"].append(data["task"])
    self.scope["session"].save()
```

6. Finally, in JavaScript, we add a `click` event to the button to send the text with the task to the Consumer class:

```javascript
/**
 * Send new task to TODO list
 * @param event
 * @return {void}
 */
function addTask(event) {
    const task = document.querySelector('#task');
    sendData({
        action: 'Add task',
        data: {
            task: task.value
        }
    }, myWebSocket);
    // Clear input
    task.value = '';
}
/**
 * Update events in every page
 * return {void}
 */
function updateEvents() {
    // Nav
    setEventsNavigation(myWebSocket);
...
    // Add task
    const addTaskButton = document.querySelector
        ('#add-task');
    if (addTaskButton !== null) {
        addTaskButton.removeEventListener('click',
            addTask, false);
        addTaskButton.addEventListener('click',
            addTask, false);
    }
}
```

We must update if new DOM elements have appeared. Otherwise, the events will stop working if the previous HTML has been deleted. The steps to follow are to stop listening to the previous events if they exist, and add the new ones. If we don't do this, the events will be lost or duplicated.

The event to be executed is simple. We capture the `#task` field and send the `Consumer` class the text of the task.

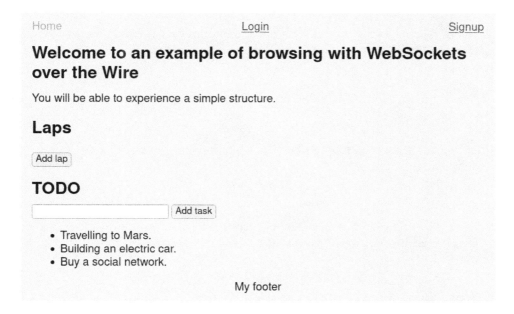

Figure 6.7 – Displaying a list of tasks from a session

We are already able to work with sessions and even create HTML from their content. Now, we just need to implement some security measures to prevent CSRF exploits.

Avoiding cross-site request forgery (CSRF) with WebSockets

By using sessions, we are exposing users to a CSRF attack unless we put appropriate measures in place.

> **CSRF attacks**
>
> CSRF attacks are malicious attacks on a website in which unauthorized commands are sent from one user to a second site with hidden forms, AJAX requests, or any other method in a hidden way.
>
> You can find a reference here: `https://en.wikipedia.org/wiki/Cross-site_request_forgery`.

Channels provides a tool that will help us to avoid this type of attack in a simple way:

1. We define the allowed Hosts in `project_template/settings.py`. In our case, we are using environment variables inside Docker:

```
ALLOWED_HOSTS = os.environ.get("ALLOWED_HOSTS").
split(",")
```

2. We edit `project_template/asgi.py`, by importing `OriginValidator`. We must pass two parameters: `URLRouter` (or any intermediary middleware) and the `Hosts` we want to protect:

```
# project_template/asgi.py
import django

os.environ.setdefault("DJANGO_SETTINGS_MODULE", "project_
template.settings")
from django.conf import settings
django.setup()
from django.core.asgi import get_asgi_application
from channels.security.websocket import OriginValidator #
New line
from channels.auth import AuthMiddlewareStack
from channels.routing import ProtocolTypeRouter,
URLRouter
from django.urls import re_path
from app.app_template.consumers import ExampleConsumer


application = ProtocolTypeRouter(
    {
        # Django's ASGI application to handle traditional
HTTP requests
        "http": get_asgi_application(),
        # WebSocket handler
        # Update
        "websocket": OriginValidator
            (AuthMiddlewareStack(
            URLRouter(
```

```
                [
            re_path(r"^ws/example/$",
                ExampleConsumer.as_asgi()),


        ), settings.ALLOWED_HOSTS)
    }
```

As this functionality is so quick to implement, it is highly recommended that it always be part of our future projects or integrated into the template that we use as a base.

Summary

In this chapter, we have added some very interesting new capabilities to our project: switching between pages, creating server-side rendering versions of each path, creating a dynamic page, modifying URLs, updating specific sections, working with sessions, and avoiding CSRF with WebSockets.

We now already have the basic skills to build a dynamic site with database access, group management, partial or full HTML rendering, event control that triggers backend actions, form creation, and some security measures. One question may be echoing in your head: was it worth all the effort? Just think that we can now create SPAs with minimal use of JavaScript, we don't need to build an API to connect the frontend and the backend, and the time between requests and their responses is ridiculously low, avoiding the use of loading in many cases. The complexity of the projects also has decreased and we can avoid the installation of several frontend libraries. Judge for yourself. The most amazing thing is that we have only used Django and Channels; the potential we can achieve by adding other Python extensions is infinite.

In the next chapter, *Chapter 7, Creating a Real-Time Blog Using Only Django*, we will put all the pieces together to exemplify a real case that we can use for our own project or an external one.

7

Creating a Real-Time Blog Using Only Django

In *Chapter 6*, *Creating SPAs on the Backends*, we learned essential features for setting up an SPA using HTML over WebSockets, such as changing pages, components, and sessions. We even went a step further by creating a server-side rendering system for each page so that search engines can index all content – a feature that didn't require much effort as we are inside Django.

We now have the skills and maturity to make applications with all the features that SPA development entails. Now is the time! We will unify all the knowledge acquired in the creation of a perfectly prepared blog. Undoubtedly, this is an excellent exercise regardless of the language or framework that we want to assimilate; it encompasses all the basic tasks of any web development: querying, filtering, and adding to a database (search engine and comments), generating HTML from results (a list of articles and an individual page), use of views (SSR), routing (static pages), processing and validating forms (incorporating a new comment), and finally, pagination.

This is an exam to prove to the world, and to yourself, that you have the basic knowledge in the subject you are learning. It can even be a good technical test.

During the creation of the blog, we will be doing the following:

- Creating models for the database
- Generating fake articles and comments
- Listing of articles
- Navigating between articles with pagination
- Adding an article search engine
- Creating a static page
- Moving between pages and generating a browser

- Implementing an individual page per article
- Adding a list of comments
- Adding new comments
- Offering an **Really Simple Syndication (RSS)** feed

During the chapter, we will work in small milestones, following an order that allows us to incorporate each element organically, without jumping from one feature to another until it is finished. You can find the code for each of the features (in the preceding list) that we will be implementing separately.

In order not to extend the example too much, we will start with a code base that we have used in previous chapters.

Technical requirements

All the code of the different sections can be found at the following link:

`https://github.com/PacktPublishing/Building-SPAs-with-Django-and-HTML-Over-the-Wire/tree/main/chapter-7`

As in other examples, I will start from the template that we built in *Chapter 4, Working with the Database*:

`https://github.com/PacktPublishing/Building-SPAs-with-Django-and-HTML-Over-the-Wire/tree/main/chapter-4/initial-template`

If you find some small differences, it is because I have made some minor adjustments. For example, I have named the project `blog`, the app `website`, and changed the path to `http://blog.localhost`, although, as always, you are free to name each element freely.

Creating models for the database

We will start by building two tables in the database: `Post`, which will contain the articles, and `Comment`, so that readers can leave their opinions next to the articles.

In `app/website/models.py`, add the following database structure:

```python
from django.db import models
from django.utils.text import slugify
from django.urls import reverse


class Post(models.Model):
    # Fields: Title of the article, name of the author,
    content of the article and date of creation.
    title = models.CharField(max_length=200, unique=True)
```

```python
    author = models.CharField(max_length=20)
    content = models.TextField()
    created_at = models.DateTimeField(auto_now_add=True)

    class Meta:
        ordering = ["-created_at"]

    @property
    def slug(self):
        return slugify(self.title)

    @property
    def summary(self):
        return self.content[:100] + "..."

    @property
    def get_absolute_url(self):
        return reverse("single post", kwargs={"slug":
            self.slug})

    def __str__(self):
        return self.title


class Comment(models.Model):
    # Fields: Name of the author, content of the comment,
    relation to the article and date of creation.
    author = models.CharField(max_length=20)
    content = models.TextField()
    post = models.ForeignKey(Post, on_delete=models.
        CASCADE)
    created_at = models.DateTimeField(auto_now_add=True)

    def __str__(self):
        return self.name
```

Let's look at the properties of `Post`:

- `slug`: We will use the title of the article to differentiate the routes. For example, if it is titled `Penguins have just conquered the world`, its final path will be `http://blog.localhost/penguins-have-just-conquered-the-world`. With this property, we get the title ready to be used for different purposes, such as feeding other properties or searching for the ID of an article.

> **Slug**
>
> Slug is a format used in URLs to make them more readable, where spaces are replaced by single dashes and text is converted to lowercase. In areas such as SEO, it is used to explain the content of the page.

- `summary`: When we list the articles, we will show a small portion of the original article. With this property, we limit the portion of the article shown to 100 characters, plus we add some nice dots at the end of the sentence. It's not perfect, as it counts spaces and doesn't check the initial length, but it's certainly sufficient for the purpose.
- `get_absolute_url`: Through the paths defined in `urls.py`, we will build the hyperlinks for each article. Why? We will move dynamically. They are for the RSS feed, for example, or a future site map.

The next step, as we have done in each activity, is to enter the Django container terminal and execute the following:

```
python3 manage.py makemigrations
python3 manage.py migrate
```

The database is ready. However, without data, it is not practical. As on other occasions, we will create fake content that simulates the final appearance of the blog.

Generating fake articles and comments

After defining the database from the models, we are going to generate random data that we will use to work more comfortably.

We create `make_fake_data.py` with the following content:

```
from app.website.models import Post, Comment
from faker import Faker


# Delete all posts and comments
```

```
Post.objects.all().delete()

# Create fake object
fake = Faker()

def get_full_name():
    return f"{fake.first_name()} {fake.last_name()}"

# Create 30 posts
for _ in range(30):
    post = Post(
        title=fake.sentence()[:200],
        content=fake.text(),
        author=get_full_name()[:20],
    )
    post.save()

# Create 150 comments
for _ in range(150):
    comment = Comment(
        author=get_full_name()[:20],
        content=fake.text(),
        post=Post.objects.order_by("?").first(),
    )
    comment.save()
```

The code we are going to run will generate random information. The steps we follow are as follows:

1. We delete all the articles, or Post. The first time we run it, there will be nothing to delete, but thereafter, it will delete everything it finds.

2. We generate 30 new articles.

3. We generate 150 comments, or Comment, and assign them to articles randomly. This way, they will be distributed irregularly, with cases where there are articles with no comments and others with a large number.

Finally, in the Django container terminal, we execute the script we have just built:

```
python3 manage.py shell < make_fake_data.py
```

Our database is populated with information. Now, we are going to focus on the logic of the blog – for example, listing all the articles in HTML.

Listing of articles

We have prepared the database through the models and by including fake information with the necessary elements, enabling us to focus on how the customer is going to visualize the content.

Before building the different pages, we will need a base for all templates. In app/website/templates/base.html, we include the main layout:

```
{% load static %}
<!doctype html>
<html lang="en">
<head>
    <meta charset="UTF-8">
    <meta name="viewport" content="width=device-width,
        user-scalable=no, initial-scale=1.0, maximum-
            scale=1.0, minimum-scale=1.0">
    <title>Example website</title>
    <link rel="stylesheet" href="{% static 'css/main.css'
        %}">
    <script defer src="{% static 'js/index.js' %}">
    </script>
</head>
<body
        data-host="{{ request.get_host }}"
        data-scheme="{{ request.scheme }}"
    >
    <div class="container">
        <header>
            <nav id="nav" class="nav">{% include
                'components/_nav.html' %}</nav>
        </header>
        <main id="main">{% include page %}</main>
```

```
        <footer class="footer">My footer</footer>
    </div>
</body>
</html>
```

We have included areas to redraw elements such as the browser, with #nav, and the main content of future pages, with #main.

Now, we are going to create the welcome page of the blog where we will list the articles.

The first step will be creating an HTML template to generate the list of the different blog articles, which will be fed by a future database query. In app/website/templates/pages/all_posts. html, we add the following code:

```
<h1> All posts</h1>
<hr>
<section>
    {# List posts #}
    <div id="all-posts">
        {% include "components/all_posts/list.html" %}
    </div>
    {# End list posts #}
</section>
```

We have separated the list of articles in a component hosted in app/website/templates/ components/all_posts/list.html because it will be useful when we do the pagination.

With the following code, let's show the list of all the articles that will be displayed inside #all-posts by means of include:

```
{% for post in posts %}
    <article>
        <header>
            <h2>{{ post.title }}</h2>
        </header>
        <p>{{ post.summary }}</p>
        <p>{{ post.author }}</p>
        <footer>
            <p>
                <a class="post-item__link" href="#" data-
```

```
                              target="single post" data-id="{{
                              post.id }}"> Read more</a>
            </p>
        </footer>
    </article>
{% endfor %}
```

At the moment, the hyperlink to go to the individual page of the article does not work. When we have the right template, we will come back to give it the logic with JavaScript. However, we have already prepared the dataset for the dynamic page change: the name of the page to load (data-target) and its ID (data-id).

In app/website/views.py, we create the following view:

```
from django.shortcuts import render
from .forms import SearchForm, CommentForm
from .models import Post, Comment


def all_posts(request):
    return render(
        request,
        "base.html",
        {
            "posts": Post.objects.all()[:5],
            "page": "pages/all_posts.html",
            "active_nav": "all posts",
        },
    )
```

We will only list the first five items; this is the number of items we will display per page.

In blog/urls.py, we assign the root of the site:

```
from django.contrib import admin
from django.urls import path
from app.website import views, feed


urlpatterns = [
```

```
        path("", views.all_posts, name="all posts"),
        path("admin/", admin.site.urls),
]
```

When you pull up Docker, via `docker-compose.yaml`, and go to `http://blog.localhost`, you will find the articles:

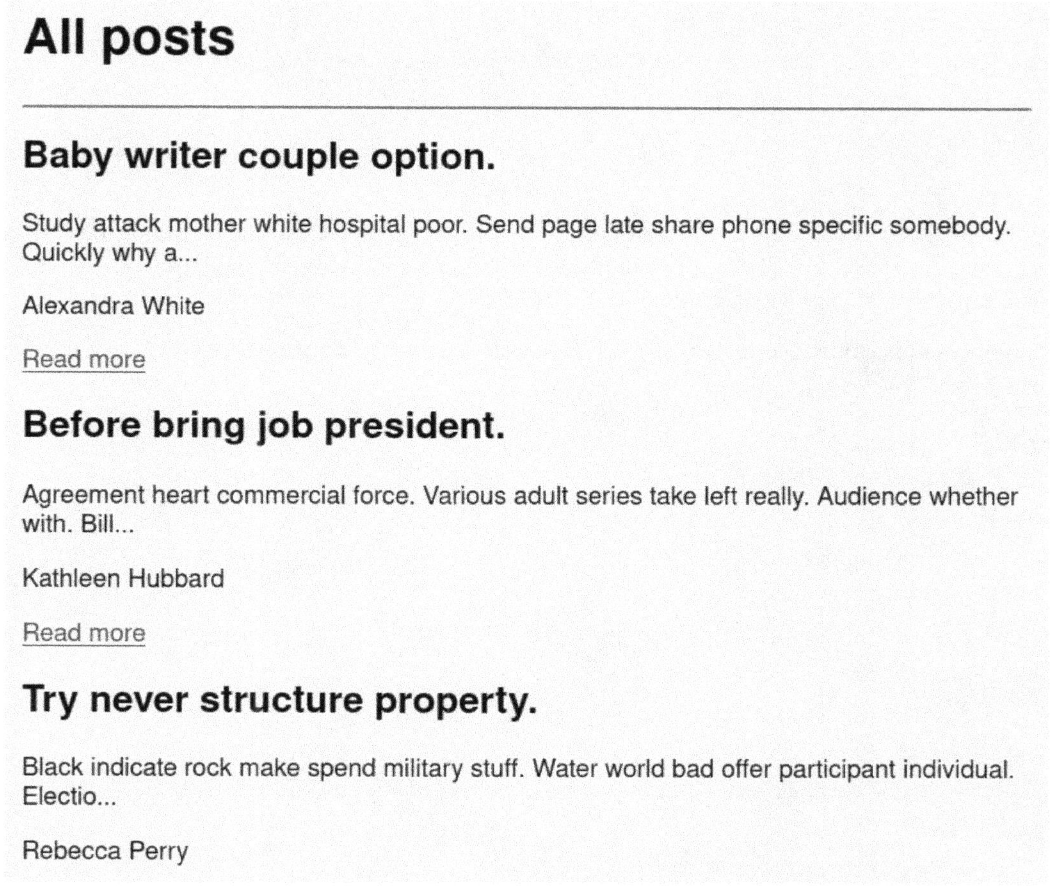

All posts

Baby writer couple option.

Study attack mother white hospital poor. Send page late share phone specific somebody. Quickly why a...

Alexandra White

Read more

Before bring job president.

Agreement heart commercial force. Various adult series take left really. Audience whether with. Bill...

Kathleen Hubbard

Read more

Try never structure property.

Black indicate rock make spend military stuff. Water world bad offer participant individual. Electio...

Rebecca Perry

Figure 7.1 – Displays the first five articles at the root of the blog

What if I want to see more articles? We can't, although we will resolve this in the next section. The next challenge will be to solve the problem by pagination, or rendering the next five posts continuously.

Navigating between articles with pagination

We are able to show visitors the latest articles, but they are unable to view previous posts. We are going to include a button that allows us to render other articles on the welcome page, and we will incorporate them in blocks of five.

We add a component with the button. In `app/website/templates/components/all_posts/_button_paginator.html`, add the following HTML:

```
{% if not is_last_page %}
<button class="button" id="paginator" data-next-page="{{
    next_page }}">More posts</button>
{% endif %}
```

We will only show the button if we are not on the last page, which we will manage with the `is_last_page` variable. In addition, we will include a dataset with the `next_page` variable to tell the backend the next page we want to render.

The component is embedded in `app/website/templates/components/all_posts/list.html`:

```
<h1>All posts</h1>
<hr>
<section>
    {# List posts #}
    ...
    {# End list posts #}
    {# Paginator #}
    <div id="paginator">
        {% include
            "components/all_posts/_button_paginator.html" %}
    </div>
    {# End paginator #}
</section>
```

After designing the visual part, we will focus on the usual flow to give the logic.

We go to `static/js/index.js` to capture the click event and send to the Consumer the `"Add next posts"` action with the number of the page we want to render.

I have omitted the lines that are already present in the template to simplify the example:

```javascript
/**
 * Event to add a next page with the pagination
 * @param event
 */
function addNextPaginator(event) {
    const nextPage = event.target.dataset.nextPage;
    sendData({
        action: "Add next posts",
        data: {
            page: nextPage
        },
    }, myWebSocket);
}


/**
 * Update events in every page
 * return {void}
 */
function updateEvents() {
    ...

    // Paginator
    const paginator = document.querySelector("#paginator");
    if (paginator !== null) {
        paginator.removeEventListener("click",
            addNextPaginator, false);
        paginator.addEventListener("click",
            addNextPaginator, false);
    }
    ...
}
```

We add to the Consumer, hosted in `app/website/consumers.py`, the appropriate call to action if we receive "Add next posts".

As we have done on several occasions, we will create a link in the Consumer class between the action required by the frontend and the function hosted in `actions.py`:

```python
from channels.generic.websocket import
    JsonWebsocketConsumer
from asgiref.sync import async_to_sync
import app.website.actions as actions


class BlogConsumer(JsonWebsocketConsumer):
    room_name = "broadcast"

    def connect(self):
        """Event when client connects"""
        # Accept the connection
        self.accept()
        # Assign the Broadcast group
        async_to_sync(self.channel_layer.group_add)
            (self.room_name, self.channel_name)

    def disconnect(self, close_code):
        """Event when client disconnects"""
        pass

    def receive_json(self, data_received):
        ...
        # Get the data
        data = data_received["data"]
        # Depending on the action we will do one task or
         another.
        match data_received["action"]:
            case "Change page":
                actions.send_page(self, data)
            case "Add next posts":
                actions.add_next_posts(self, data)

    def send_html(self, event):
        ...
```

In app/website/actions.py, we declare the add_next_posts function:

```python
POST_PER_PAGE = 5

def add_next_posts(self, data={}):
    """Add next posts from pagination"""
    # Prepare context data for page
    page = int(data["page"]) if "page" in data else 1
    start_of_slice = (page - 1) * POST_PER_PAGE
    end_of_slice = start_of_slice + POST_PER_PAGE
    context = {
        "posts": Post.objects.all()[start_of_slice:end_of_
slice],
        "next_page": page + 1,
        "is_last_page": (Post.objects.count() //
            POST_PER_PAGE) == page,
    }


    # Add and render HTML with new posts
    self.send_html(
        {
            "selector": "#all-posts",
            "html": render_to_string
                ("components/all_posts/list.html", context),
            "append: True,
        }


    # Update paginator
    self.send_html(
        {
            "selector": "#paginator",
            "html": render_to_string(
                "components/all_posts/_button_paginator.
                    html", context
            ),
        }
```

We are undertaking a number of important actions:

- We're saving the page to display. If it is not provided, we assume it is the first one.

- We're calculating the initial and final cut-off of the results.

- We're carrying out the query.

- We're calculating what the next page will be – the current page plus one.

- We're checking whether we are on the last page. It will be important to know whether we should print the paging button or not.

- We're rendering new articles and adding them to #all-posts.

- We're redrawing the paging button, as it needs to store what the next page is and hide it if there are no more articles.

There is only one detail left. Give the initial parameters to the view (app/website/views.py):

```python
def all_posts(request):
    return render(
        request,
        "base.html",
        {
            "posts": Post.objects.all()[:5],
            "page": "pages/all_posts.html",
            "active_nav": "all posts",
            "next_page": 2, # New
            "is_last_page": (Post.objects.count() //
                POST_PER_PAGE) == 2, # New
        },
```

We can now start rendering new results:

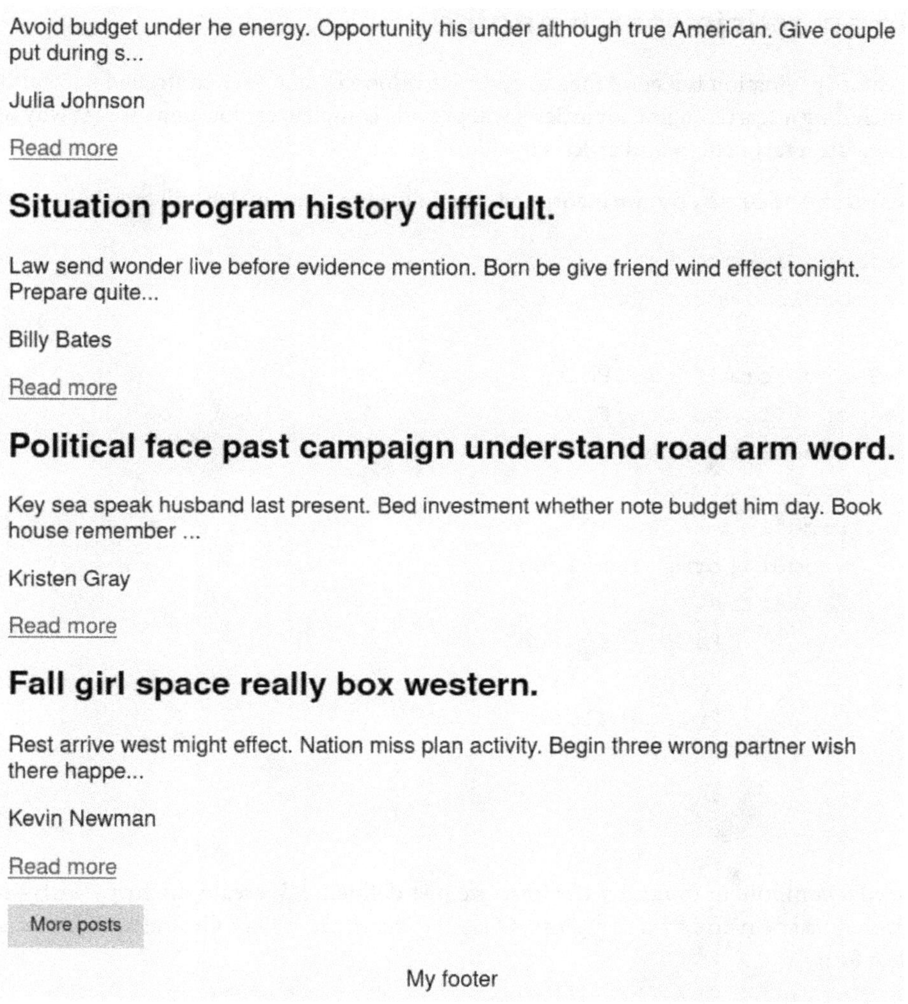

Avoid budget under he energy. Opportunity his under although true American. Give couple put during s...

Julia Johnson

Read more

Situation program history difficult.

Law send wonder live before evidence mention. Born be give friend wind effect tonight. Prepare quite...

Billy Bates

Read more

Political face past campaign understand road arm word.

Key sea speak husband last present. Bed investment whether note budget him day. Book house remember ...

Kristen Gray

Read more

Fall girl space really box western.

Rest arrive west might effect. Nation miss plan activity. Begin three wrong partner wish there happe...

Kevin Newman

Read more

More posts

My footer

Figure 7.2 – The pagination of articles

The experience would be more pleasant with a nice animation or delay; it is so fast loading that a visitor may not notice the new elements. We can leave an issue to the future web designer. Our task is not finished yet – what if the visitor is looking for a specific article? The pagination becomes cumbersome; everything would be easier with a simple search engine.

Adding an article search engine

Offering visitors pagination is a good idea to optimize resources and offer controlled navigation. In addition, including a search engine for articles will provide complete exploration. That is why we are going to integrate a text field to find articles by title.

In `app/website/forms.py`, we incorporate the following form, which will only have one field:

```
from django import forms
from . models import Comment


class SearchForm(forms.Form):
    search = forms.CharField(
        label="Search",
        max_length=255,
        required=False,
        widget=forms.TextInput(
            attrs={
                "id": "search",
                "class": "input",
                "placeholder": "Title...",
            }
        ),
```

We will need a component to render the form we just defined. We create the `app/website/templates/components/all_posts/form_search.html` file and add the `search` field inside a form:

```
<form id="search-form" action="">
    {{ form.search }}
    <input class="button" type="submit" value="Search">
</form>
```

In the article listing page, `app/website/templates/pages/all_posts.html`, we include the `search` component:

```
<h1> All posts</h1>
<hr>
```

```
{# Search #}
<section id="form-search">
    {% include "components/all_posts/form_search.html" %}
</section>
{# End search #}
<hr>
<section>
    {# List posts #}
    ...
    {# End list posts #}
    {# Paginator #}
    ...
    {# End paginator #}
</section>
```

Don't forget to include it in the view (app/website/views.py):

```
def all_posts(request):
    return render(
        request,
        "base.html",
        {
            "posts": Post.objects.all()[:5],
            "page": "pages/all_posts.html",
            "active_nav": "all posts",
            "form": SearchForm(), # New
            "next_page": 2,
            "is_last_page": (Post.objects.count() //
                POST_PER_PAGE) == 2,
        },
```

We will see the form when the page loads, although for the moment it is decorative, as there is no logic behind it:

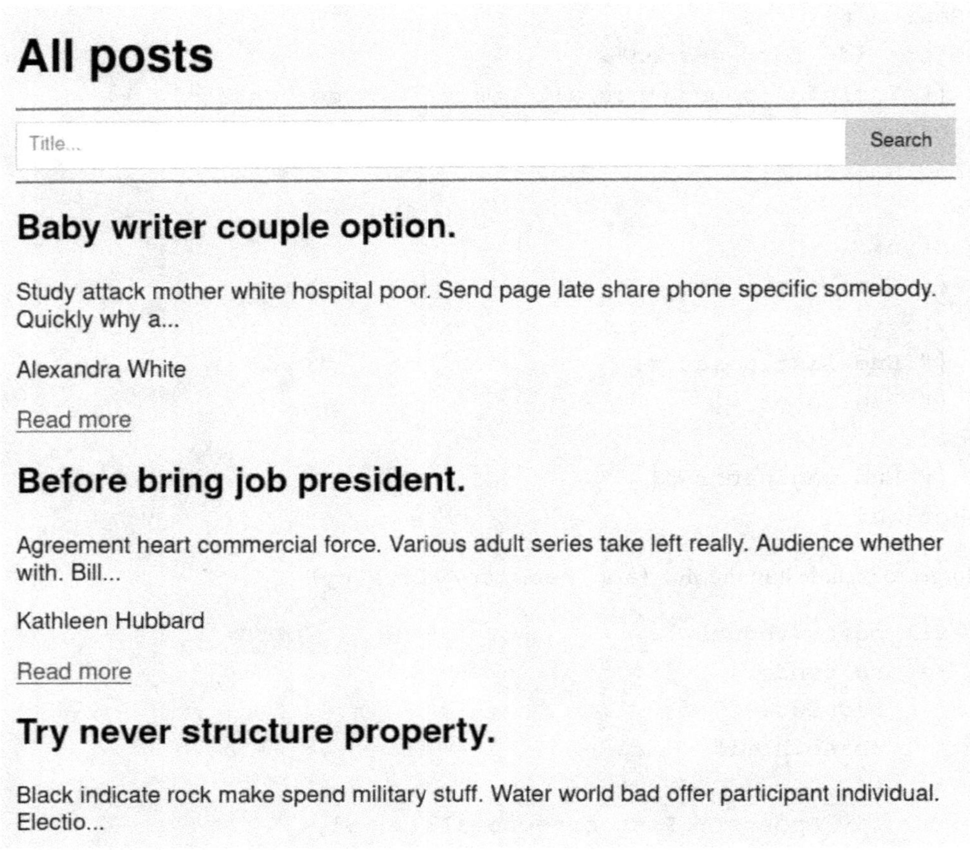

All posts

Title...	Search

Baby writer couple option.

Study attack mother white hospital poor. Send page late share phone specific somebody. Quickly why a...

Alexandra White

Read more

Before bring job president.

Agreement heart commercial force. Various adult series take left really. Audience whether with. Bill...

Kathleen Hubbard

Read more

Try never structure property.

Black indicate rock make spend military stuff. Water world bad offer participant individual. Electio...

Figure 7.3 – Displaying the browser

Now, let's go to `static/js/index.js` to make it work. We'll capture the form's submit event and send the Consumer the `"Search"` action with the text to search for:

```
/**
 * Event to request a search
 * @param event
 */
function search(event) {
    event.preventDefault();
    const search = event.target.querySelector("#search").
value;
    sendData({
```

```
        action: "Search",
        data: {
            search: search
        },
    }, myWebSocket);
}
/**
 * Update events in every page
 * return {void}
 */
function updateEvents() {
...

    // Search form
    const searchForm = document.querySelector("#search-
        form");
    if (searchForm !== null) {
        searchForm.removeEventListener("submit", search,
            false);
        searchForm.addEventListener("submit", search,
            false);
    }
...
}
```

The frontend has already sent us the request and the information we need. Now, the Consumer (app/ website/consumers.py) should execute the appropriate action:

```
match data_received["action"]:
...

            case "Search":
                actions.search(self, data)
...
```

Then, in the action (app/website/actions.py), we include the search function:

```
def search(self, data={}):
    "Search for posts"        ""
```

```
# Prepare context data for page
context = {
    "posts": Post.objects.filter
        (title__icontains=data["search"])
            [:POST_PER_PAGE].
}

# Render HTML page and send to client
self.send_html(
    {
        "selector": "#all-posts",
        "html": render_to_string
            ("components/all_posts/list.html", context),
    }
)
```

As you can see, the code is simple. All we do is filter the database by getting all articles containing data["search"], ignoring uppercase or lowercase text (icontains). We also limit the results to five articles.

That's it. We can search and dynamically display the results:

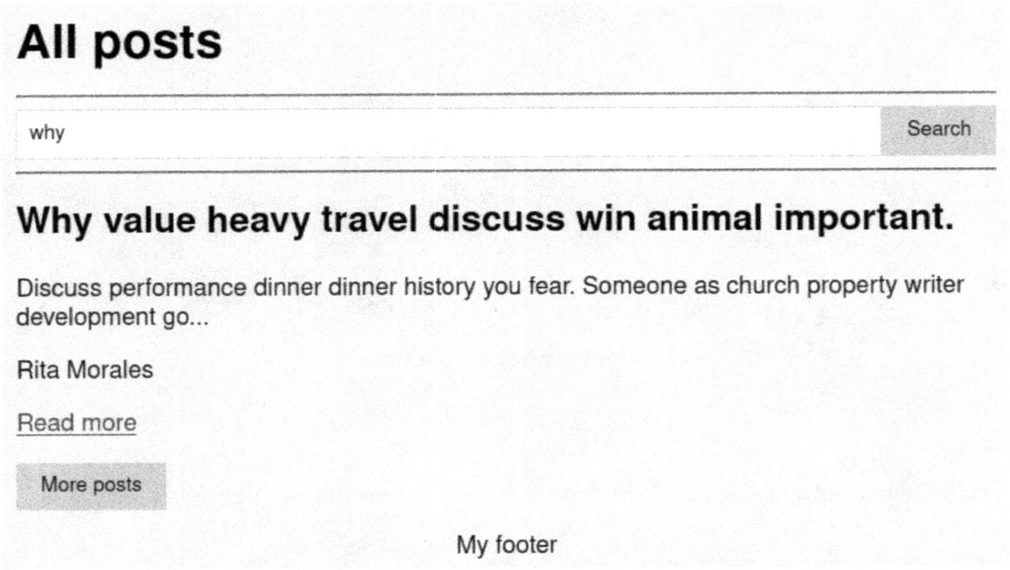

All posts

why   Search

Why value heavy travel discuss win animal important.

Discuss performance dinner dinner history you fear. Someone as church property writer development go...

Rita Morales

Read more

More posts

My footer

Figure 7.4 – Shows the results of a search for the word "why"

If you search by leaving the string empty, you will return to the previous state where the items are listed without filtering.

The next point to discuss is the navigation between pages. For this, we will create a static page, where we can describe the blog or or display the **About us** page, and a navigator to move between the existing pages.

Creating a static page

We are in a situation where we need to grow with new pages to split logic and HTML structures. The first step will be to create a static page.

We create app/website/templates/pages/about_us.html with simple text:

```
<h1> About us</h1>
<p> Lorem ipsum dolor sit amet, consectetur adipisicing elit.
Ad animi aut beatae commodi consectetur cumque ipsam iste
labore laudantium magni molestiae nobis nulla quod quos tempore
totam velit, voluptas voluptates!</p>
```

We edit the views (app/website/views.py), including about:

```
def about(request):
    return render(
        request,
        "base.html",
        { "page": "pages/about_us.html", "active_nav":
            "about us"},
    )
```

Then, we give it a path in blog/urls.py:

```
urlpatterns = [
...
        path("about-us/", views.about, name="about us"),
...
```

We can now go to http://blog.localhost/about-us/ to view the page:

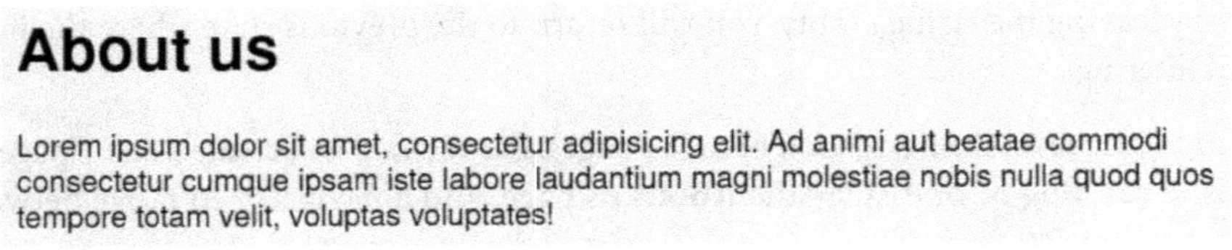

About us

Lorem ipsum dolor sit amet, consectetur adipisicing elit. Ad animi aut beatae commodi consectetur cumque ipsam iste labore laudantium magni molestiae nobis nulla quod quos tempore totam velit, voluptas voluptates!

Figure 7.5 – Rendering the About us page

I totally agree with you; this section hasn't been very... I plead guilty! Creating a static page is the most basic thing we can do in Django. Now, it's time for the hard part: dynamically scrolling between pages and creating a browser.

Moving between pages and generating a browser

Visitors need to navigate between different pages; a simple button structure and corresponding logic for loading the appropriate templates will need to be incorporated.

We are going to create a browser to dynamically jump between pages or, in other words, request the backend to render the page in the right place:

1. The first step is to create a component with hyperlinks. We create a file in `app/website/templates/components/_nav.html` with the following structure:

```
<ul class="nav__ul">
    <li>
        <a
                href="#"
                class="nav__link nav__link nav__link--
page{% if active_nav == "all posts" %} active{% endif
%}""
                data-target="all posts"
        >
            All posts
        </a>
    </li>
    <li>
        <a
                href="#"
                class="nav__link nav__link nav__link
                page{% if active_nav == "about us"
```

```
                                %} active{% endif %}"
                    data-target="about us"
          >
                About us
          </a>
     </li>
</ul>
```

The `active_nav` variable is worth mentioning. We have defined it in each view for this particular component. It will add a CSS class that visually marks where the visitor is. We also include the `target` dataset to know where each hyperlink should point to.

Next, we are going to capture the events of the hyperlinks in JavaScript whose objective is to change the page, both those present in the browser and the list of articles:

2.   We add the following in `static/js/index.js`:

```
/**
* Send message to update page
* @param {Event} event
* @returns {void}
*/
function changePage(event) {
    event.preventDefault();
    sendData({
        action: "Change page",
        data: {
            page: event.target.dataset.target,
            id: event.target. dataset?.id
        }
    }, myWebSocket);
}


/**
* Update events in every page
* return {void}
*/
function updateEvents() {
```

```javascript
. . .
    // Navigation
    document.querySelectorAll(".nav__link--page").
forEach(link => {
        link.removeEventListener("click", changePage,
false);
        link.addEventListener("click", changePage,
false);
    });
    // Link to single post
    const linksPostItem = document.querySelectorAll
        (".post-item__link");
    if (linksPostItem !== null) {
        linksPostItem.forEach(link => {
            link.removeEventListener("click",
                changePage, false);
            link.addEventListener("click", changePage,
                false);
        });
    }
...
}
```

When the hyperlinks are clicked, a request will be made to the Consumer to change the page by sending the name of the template and, if it exists, the ID of the post.

3. We include in the Consumer (app/website/consumers.py) the send_page call when we receive "Change page":

```python
. . .
ca"e "Change page":
actions.send_page(self, data)
. . .
```

4. In action, we edit the send_page function, as we have done in previous projects, by adding the template context template:

```
POST_PER_PAGE = 5

def send_page(self, data={}):
...

    match page:
        case "all posts":
            context = {
                "posts": Post.objects.all()
                    [:POST_PER_PAGE],
                "form": SearchForm(),
                "next_page": 2,
                "is_last_page": (Post.objects.count()
                // POST_PER_PAGE) == 2,
            }

    ...
```

It is practically the same as the view in charge of displaying all items.

We can now move between pages and visualize where we are with the CSS styling:

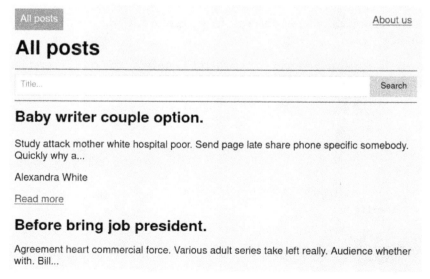

Figure 7.6 – The browser shows that we are in All posts

We manage all cases. The visitor can navigate from any type of page, from a dynamic page to another where the content is static.

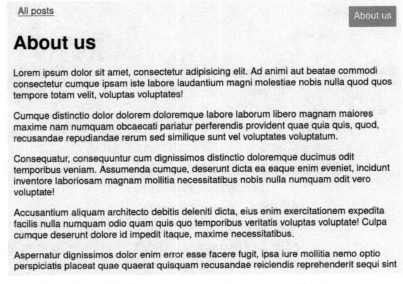

Figure 7.7 – The browser shows that we are in About us

However, the dynamic page showing the full text of the article and comments is still missing. By having a system for navigation, it will be relatively easy to incorporate it.

Implementing an individual page per article

We have the opportunity to create a page that renders an entire article, which will be the basis for the entire commenting system.

We create the template in `app/website/templates/pages/single_post.html` with basic but sufficient HTML for the minimum `Post` fields:

```
<section>
    {# Post #}
    <article>
        <header>
            <h1>{{ post.title }}</h1>
        </header>
        <div>{{ post.content }}</div>
        <footer>
```

```
            <p>{{ post.author }}</p>
        </footer>
    </article>
    {# End post #}
</section>
```

Now, we are going to focus on server-side rendering by creating the view and its path.

In app/website/views.py, we include the following function:

```
from .models import Post
def single_post(request, slug):
    post = list(filter(lambda post: post.slug == slug, Post.
objects.all()))[0]
    return render(
        request,
        "base.html",
        {
            "post: post,
            "page": "pages/single_post.html",
        },
```

Why use the filter function? As we have decided that the URL will be formed by a slug, when we receive the request to render the view, we will need to look for the post with the slug property. Django does not allow you to perform a query through a property. In other words, we will have to perform manual filtering.

We incorporate the route in blog/urls.py:

```
urlpatterns = [
    ...
    path("article/<slug:slug>/", views.single_post,
        name="single post"),
    ...
```

Now, we need to incorporate a context, or a set of variables needed to render the new HTML, when the frontend request to switch pages. In `app/website/actions.py`, we add the following:

```python
data_reverse = {}
match page:
    ...

        case "single post":
            post = Post.objects.get(id=data["id"])
            context = {
                "post: post,
            }
            data_reverse = {"slug": post.slug}
    ...
self.send_html(
        {
            "selector": "#main",
            "html": render_to_string(f
                "pages/{template_page}.html", context),
            "url": reverse(page, kwargs=data_reverse),
        }
```

At this moment, from the list of articles, we will be able to load the single template. Also, the path will change with the URL structure of the slug:

All posts   About us

Try never structure property.

Black indicate rock make spend military stuff. Water world bad offer participant individual. Election hotel lose center situation everything baby.

Rebecca Perry

My footer

Figure 7.8 – The individual article page is rendered

However, the page is not finished; we still need to list the comments at the bottom of the template and include a form to add new ones.

Adding a list of comments

The blog is functional: we can list articles, navigate between pages, paginate, and perform a search. But an essential element is still missing: comments. That's why we are going to print all the comments that belong to an article.

We start by creating a template that lists all the comments. We add a new component in `app/website/templates/components/_list_of_comments.html` with the following content:

```
{% for comment in comments %}
    {% include "components/_single_comment.html" with
        comment=comment %}
{% endfor %}
```

This, in turn, will need the `app/website/templates/components/_single_comment.html` component:

```
<article>
    <h2>{{ comment.author }}</h2>
    <p>{{ comment.content }}</p>
    <p>{{ comment.created_at }}</p>
</article>
```

In the views (`app/website/views.py`), we make a query with all the comments that belong to the post we are viewing and send it to the template:

```
from .models import Post, Comment
def single_post(request, slug):
    ...
        {
            "post: post,
            "page": "pages/single_post.html",
            "active_nav": "single post",
            "comments": Comment.objects.filter(post=post), #
New
        },
```

...

We can now display a list of comments on the article.

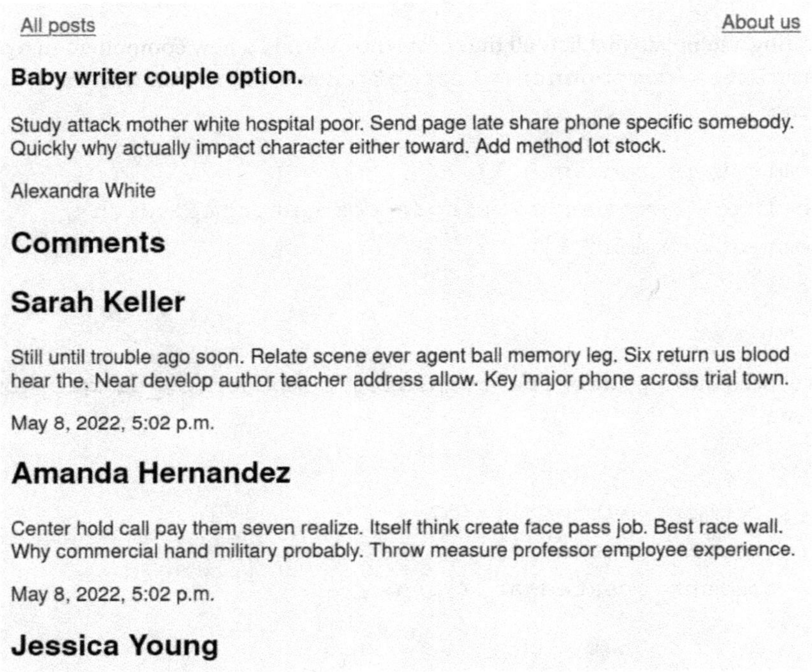

All posts   About us

Baby writer couple option.

Study attack mother white hospital poor. Send page late share phone specific somebody.
Quickly why actually impact character either toward. Add method lot stock.

Alexandra White

Comments

Sarah Keller

Still until trouble ago soon. Relate scene ever agent ball memory leg. Six return us blood
hear the. Near develop author teacher address allow. Key major phone across trial town.

May 8, 2022, 5:02 p.m.

Amanda Hernandez

Center hold call pay them seven realize. Itself think create face pass job. Best race wall.
Why commercial hand military probably. Throw measure professor employee experience.

May 8, 2022, 5:02 p.m.

Jessica Young

Figure 7.9 – All comments are rendered on the article page

However, to also display comments when we dynamically change pages, we must include the comment
variable inside send_page in actions:

```
def send_page(self, data={}):
    ...
    case "single post":
        post = Post.objects.get(id=data["id"])
        context = {
            "post: post,
            "form": CommentForm(),
            "comments": Comment.objects.filter(post=post),
# New
```

```
        }
        data_reverse = {"slug": post.slug}
...
```

We will now incorporate a form for visitors to add comments. But that's not all: we have generated random information, article listings, a single page per article, a system for dynamically switching between pages, a browser, a search engine, and a server-side rendering system. At the moment, we have a very interesting blog. Next, we'll see how to add new comments.

Adding new comments

If all the comments were written by us, it would be a bit immoral. We're going to incorporate a form so that anyone reading the article can leave a personal opinion. If you don't like what they say, you can always "manage" it with Django's admin panel. But for now, let's not be tricky; let's focus on the more technical side.

First, we add the following form in app/website/forms.py:

```python
class CommentForm(forms.ModelForm):

    author = forms.CharField(
        widget=forms.TextInput(
            attrs={
                "id": "author",
                "class": "input",
                "placeholder": "Your name...",
            }
        ),


    content = forms.CharField(
        widget=forms.Textarea(
            attrs={
                "id": "content",
                "class": "input",
                "placeholder": "Your comment...",
            }
        ),
```

```
class Meta:
    model = Comment
    fields = ("author", "content", "post")
```

There is an important difference with respect to the search engine form: we use ModelForm. Now we can create a new comment from the form object itself.

In the views (app/website/views.py), we import and include the form object in the template:

```
from . forms import SearchForm, CommentForm
def single_post(request, slug):
    ...
        {
            "post: post,
            "page": "pages/single_post.html",
            "active_nav": "single post",
            "comments": Comment.objects.filter(post=post),
            "form": CommentForm(), # New
        },

    ...
```

Now, in app/website/templates/pages/single_post.html, we render the form:

```
    ...
{# Comments #}
    <div id="comments">
        <h2> Comments</h2>
        <form id="comment-form" action="" data-post-id="{{
            post.id }}">
            {{ form.author }}
            {{ form.content }}
            <input class="button" type="submit"
                value="Add">
        </form>
        <div id="list-of-comments">
            {% include "components/_list_of_comments.html" %}
```

```
            </div>
        </div>
        {# End comments #}
    </section>
```

By clicking on any item, you will be able to view the form:

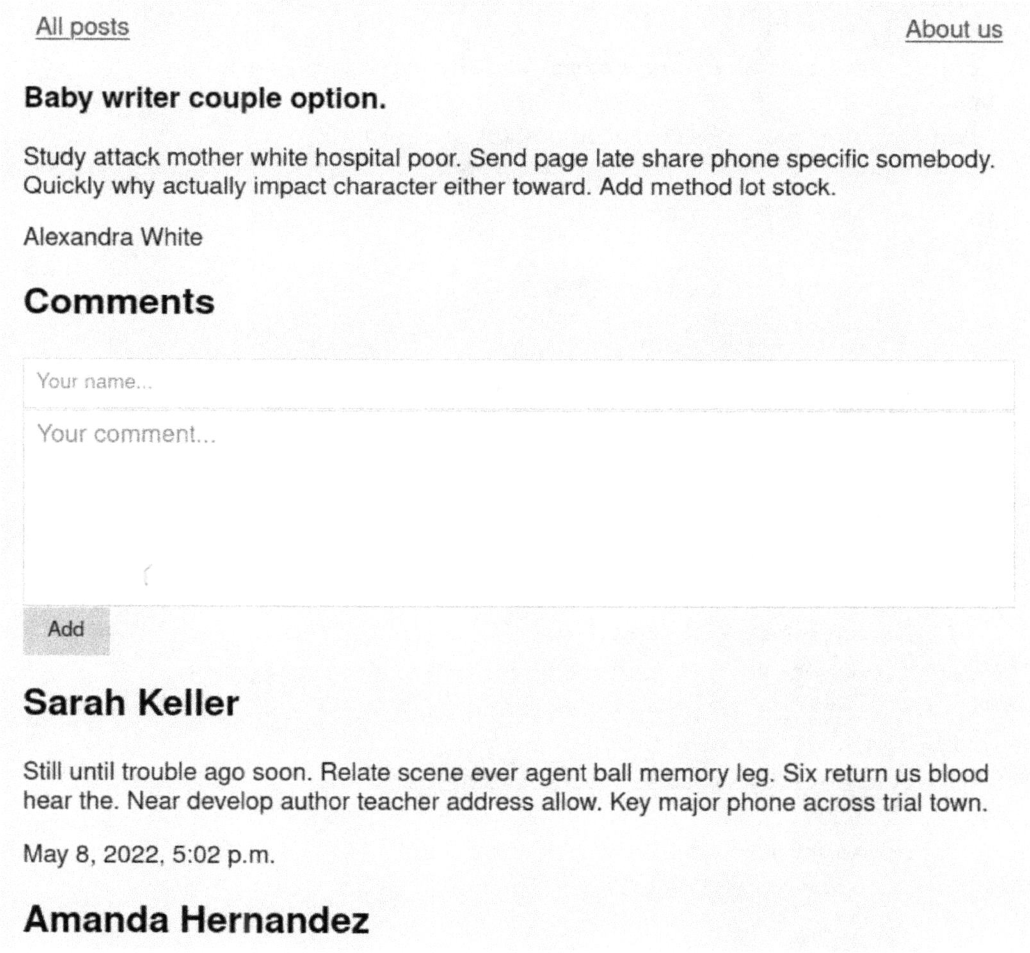

All posts   About us

Baby writer couple option.

Study attack mother white hospital poor. Send page late share phone specific somebody.
Quickly why actually impact character either toward. Add method lot stock.

Alexandra White

Comments

Your name...

Your comment...

Add

Sarah Keller

Still until trouble ago soon. Relate scene ever agent ball memory leg. Six return us blood
hear the. Near develop author teacher address allow. Key major phone across trial town.

May 8, 2022, 5:02 p.m.

Amanda Hernandez

Figure 7.10 – Render the form to add new comments

Now, we are going to process the form from the frontend. We capture the submit event and when it fires, we will get the three fields: author, content, and the article ID. We will send a request to execute the "Add comment" action.

We add in static/js/index.js the following functions:

```js
function addComment(event) {
    event.preventDefault();
    const author = event.target.querySelector("#author").
value;
    const content = event.target.querySelector("#content").
value;
    const postId = event.target.dataset.postId;
    sendData({
        action: "Add comment",
        data: {
            author: author,
            content: content,
            post_id: postId
        },
    }, myWebSocket);
}


function updateEvents() {
    ...
    // Comment form
    const commentForm = document.querySelector("#comment-
form");
    if (commentForm !== null) {
        commentForm.removeEventListener("submit", addComment,
false);
        commentForm.addEventListener("submit", addComment,
false);
    }
    ...
}
```

In the Consumer, app/website/consumers.py, we call the add_comment function inside actions if we receive "Add comment":

```
match data_received["action"]:
...

    case "Add comment":
        actions.add_comment(self, data)
```

To finish the flow, in actions (app/website/actions.py), we create the function that invokes the Consumer – add_comment:

```
from . models import Post, Comment
from . forms import SearchForm, CommentForm


def add_comment(self, data):
    """Add new comment to database"""
    # Add post
    data_with_post = data.copy()
    post = Post.objects.get(id=data["post_id"])
    data_with_post["post"] = post
    # Set initial values by CommentForm
    form = CommentForm(data_with_post)
    # Check if form is valid
    if form.is_valid():
        # Save comment
        form.save()
        # Render HTML with new comment to all clients
        async_to_sync(self.channel_layer.group_send)(
            self.room_name,
            {
                "type": "send.html", # Run "send_html()"
                    method
                "selector": "#comments",
                "html": render_to_string(
                    "components/_single_comment.html",
```

```
                    {"comment": data}, {"comment": data}.
        ),
        "append": True,
        "broadcast: True,
        "url": reverse("single post",
            kwargs={"slug": post.slug}),
    },
```

We are carrying out a set of actions that must be in the following order:

1. We obtain the post from the ID received.

2. We include the post inside the dictionary with all the information. We need to add the object to perform the validation of the form.

3. With the dictionary, we initialize the form.

4. We validate that the fields are correct. If they are not, the rest of the code will simply be ignored.

5. If they are correct, we create the new comment in the database with `form.save()`. The form knows which model to create because, internally, it is `ModelForm`, and we tell it in `forms.py`.

6. We send to all connected clients the HTML of the new comment.

7. Not only is the comment validated and saved, but it is also sent to all readers of the article in real time. However, you should be aware that we are not giving feedback in case the fields are not valid. Simply, until all fields are filled in, the information will not be processed.

We could stop here, but there is still one detail that I think is indispensable if we create a blog: an RSS feed so that our future visitors can be informed of the latest news.

Offering an RSS feed

Tech blogs are often consumed by robots, in particular by feed readers. If we want to build a feed in Django, it's really convenient. Django incorporates a framework called **Syndication** that automates tasks such as dynamic generation of XML, fields, and caching.

In app/website/feed.py, we add the following class that inherits from Feed:

```
from django.contrib.syndication.views import Feed
from django.urls import reverse
from .models import Post
```

```
class LatestEntriesFeed(Feed):
    title = "My blog"
    link = "/feed/"
    description = "Updates to posts."

    def items(self):
        return Post.objects.all()[:5]

    def item_title(self, item):
        return item.title

    def item_description(self, item):
        return item.summary

    def item_link(self, item):
        return reverse("single post", kwargs={"slug":
            item.slug})
```

Finally, we include its path in blog/urls.py:

```
...
from app.website import views, feed

urlpatterns = [
    ...
    path("feed/", feed.LatestEntriesFeed(), name="feed"),
    ...
```

You can give your favorite feed reader client the http://blog.localhost/feed/ path. If you enter it directly from the browser, an XML file will be downloaded.

Summary

We could consider this chapter as a consummation of all the skills acquired throughout the book. Not only are we able to incorporate a WebSockets server into Django, through channels; we now also have techniques to create a real-time, single-page application using Python. We now have a deep knowledge that matches the results we can achieve with other similar projects, such as LiveView in Phoenix (the most popular framework in the Elixir ecosystem), StimulusReflex, Turbo, Action Cable, or Hotwire in Ruby on Rails.

If we are looking to abstract part of the process, there are some frameworks within Django that can be useful, such as Django Sockpuppet or Django Reactor. Unfortunately, neither of them is receiving updates, although it is a great idea to find out how they are constructed in order to further expand our knowledge.

Although the backend is covered, it is still cumbersome to work with the frontend. Events have to be redeclared on every draw, and there are tasks that we repeat recurrently on every element we want to manage. We need to simplify the process.

In the next chapter, *Chapter 8*, *Simplifying the Frontend*, we will use a JavaScript event library specially designed to rebuild the DOM without altering the way it works.

Part 4: Simplifying the frontend with Stimulus

In this part, we will use one of the most widely used JavaScript frameworks to easily handle events received from the backend: Stimulus. Not only will it reduce the amount of JavaScript code, but it will also allow the frontend to define the events.

In this part, we will cover the following chapters:

- *Chapter 8, Simplifying the Frontend*

8

Simplifying the Frontend

Throughout the chapters (e.g., doing the chat project or the blog), we wrote sloppy JavaScript code. We were forced to repeat tasks every time the backend sent new HTML, cleaning up orphaned events and reassigning new ones to the newly created DOM. Our ambitions with the frontend have been quite modest. We've limited ourselves to surviving by focusing all our energies on the Django code. If we had had a tool to handle events via HTML rendered by the server, the JavaScript code would have been less verbose and much easier to work with. It's time to refactor the frontend, but we need help to do that.

Stimulus is ideal for the job. We are talking about a framework whose objective is to constantly monitor changes in the page by connecting attributes and events with functions that we indicate. We can create controllers that we will assign through datasets to the inputs or any other element that we need to incorporate an event. And, in turn, we will associate each event to some logic in JavaScript. A fantastic definition that is in Stimulus' own documentation: you should think of Stimulus as a CSS class that connects an HTML element with a set of styles.

In this chapter, we will focus on creating a minimal example with Stimulus that will serve as a basis for understanding how it works and that can be implemented in any event on the website. In order, we will cover the following points:

- Installing and configuring Stimulus

- Defining a controller

- Managing events with actions

- Capturing references with targets

- Building an application that converts text into uppercase letters

The final goal is to build a small application where we write in a text box and in real time, we visualize the same string but in capital letters. For this, we will use Stimulus to capture the event and the input value and communicate with the consumer. You will be surprised at how elegant the frontend will become when everything is in place.

Technical requirements

The example is based on the template we used in *Chapter 4, Working with the Database*:

`https://github.com/PacktPublishing/Building-SPAs-with-Django-and-HTML-Over-the-Wire/tree/main/chapter-4/initial-template`

The finished code can be found in the following repository:

`https://github.com/PacktPublishing/Building-SPAs-with-Django-and-HTML-Over-the-Wire/tree/main/chapter-8`

It is also recommended that you visit Stimulus' own documentation to learn more about important concepts such as controllers, actions, and targets:

`https://stimulus.hotwired.dev/handbook/`

And, optionally, it's recommended that you have a modern version of Node.js along with the latest version of npm. We will use it to install the Stimulus package, but we can also use the CDN.

> **CDN or Content delivery network**
>
> A CDN is a group of servers located around the world that work together to deliver content to users quickly and efficiently. It is used with static content such as images, CSS and JavaScript.

With the resources clear, we can now start implementing a better version of the frontend. We will start by installing Stimulus and talking about its different configurations.

Installing and configuring Stimulus

Before you can use Stimulus, you will need to download and install the framework. If you don't want to complicate things, you can import it from its CDN. Just add the following script in the HTML template:

```
<script type="module" src=
https://unpkg.com/@hotwired/stimulus@3.0.1/dist/stimulus.js
>
```

If you opt for this solution, you can ignore the rest of this section.

However, if you want to download Stimulus, which is very good practice, please note that it is available in the npm packages, so let's install it with a command:

```
npm i @hotwired/stimulus
```

From here, you have three different configuration possibilities: using Webpack, using another build system, or using the native JavaScript module system. We are going to focus on the last option, using modules, to simplify your implementation and not add more complexity:

1. Copy the Stimulus file to a folder inside `static`, for example, in the `static/js/vendors/` path:

    ```
    mkdir -p static/js/vendors/
    cp node_modules/@hotwired/stimulus/dist/stimulus. Js
    static/js/vendors/
    ```

2. We create a JavaScript file called `main.js` that will contain all future frontend logic and imports (including Stimulus):

    ```
    touch static/js/main.js
    ```

3. Inside the file we just created, `main.js`, we will import Stimulus and run it:

    ```
    import { Application } from "./vendors/stimulus.js";
    window.Stimulus = Application.start();
    ```

4. Finally, we import the JavaScript module into a script that will be present in the main HTML template of the application so that the browser can load it:

    ```
    <script defer type="module"
      src="{% static 'js/main.js' %}"></script>
    ```

Stimulus is ready! It's up and running and waiting for our events.

The best way to understand all the elementary concepts is to create a simple application. As we said in the introduction, we are going to build an app that has a basic functionality: convert some text from lowercase to uppercase. We will have an input and a button; when pressed, the button will show the uppercase text at the bottom.

To achieve the objective, we will learn about the three basic pillars of Stimulus: **controllers**, **actions** (not to be confused with those created in the backend), and **targets**. We will start by looking at controllers and their importance in organizing logic.

Defining a controller

The purpose of the controller is to connect the DOM with JavaScript. It will bind the inputs to a variable and the events that we indicate to a function created inside the controller.

The structure is as follows:

```
import { Controller } from "../vendors/stimulus.js".

export default class extends Controller {

    // Variables linked to inputs.
    static targets = [ "input1" ]

    // Constructor or function to be executed when the
    // controller is loaded.
    connect() {

    }

    // Simple function
    myFunction(event) {

    }
}
```

We have imported the `Controller` class that belongs to the framework itself with a combination of `import` and `from`. Then, we created a class that extends `Controller` and is also accessible from an import (`export default`). Inside, we have an example of a target called `input1` and two functions: `connect()` will be executed when Stimulus is ready and `myFunction()` is an example function that can be executed.

For the application, we will create a file in `static/js/controllers/transformer_controller.js` with the following content:

```
import { Controller } from "../vendors/stimulus.js"

export default class extends Controller {

  static targets = [ "myText" ]

    connect() {
      // Connect to the WebSocket server
        this.myWebSocket = new WebSocket(
```

```
                'ws://hello.localhost/ws/example/');
        // Listen for messages from the server
        this.myWebSocket.addEventListener("message",
                                          (event) => {
            // Parse the data received
            const data = JSON.parse(event.data);
            // Renders the HTML received from the Consumer
            const newFragment = document.createRange().
               createContextualFragment(data.html);
            document.querySelector(data.selector).
               replaceChildren(newFragment);
        });
    }

    lowercaseToUppercase(event) {
      event.preventDefault()
      // Prepare the information we will send
      const data = {
          "action": "text in capital letters",
          "data": {
              "text": this.myTextTarget.value
          }
      };
      // Send the data to the server
      this.myWebSocket.send(JSON.stringify(data));
    }
  }
}
```

As you can see, it is a reorganization of the code we used in the frontend during the previous chapters. Let's take a closer look at each part:

- In targets, we define a variable called myText that will be linked later, in the *Capturing references with targets* section, where we get the value of the input. Inside the controller, we can use the input with this.mytextTarget. A target contains all the elements of an input, such as value.

- connect() is a function that is executed when the driver is fully mounted. It is a good place to connect to the WebSockets server and set a listener event for messages.

- `lowercaseToUppercase(event)` is a function that sends the backend the text to convert to uppercase. In the next section, *Managing events with actions*, we will link the button click event to the function. For now, we just declare its logic.

After declaring the controller, we need to register it in Stimulus and give it a name. We edit `static/js/main.js` with the following code.

```
import { Application } from "./vendors/stimulus.js";
import TransformerController from
  "./controllers/transformer_controller.js"; // New line

window.Stimulus = Application.start();
// New line
Stimulus.register("transformer", TransformerController);
```

Basically, we have imported the `TransformerController` class and registered it in Stimulus with the alias `transformer`.

At the moment, Stimulus already has a controller registered, but it doesn't know which area of the DOM it should be watching and where to apply it. Let's take care of that.

In a new template, for example with the name `index.html`, we are going to create a simple form and an element to render everything coming from the backend:

```
<main>
<form>
<input type="text" placeholder="Enter text">
<input type="button" value="Transform">
</form>
<div id="results"></div>
</main>
```

The form has a field to write the text and a button that will execute the action. On the other hand, we have included a HTML `div` tag with the ID `results`, which will be the place to show the text already converted into uppercase processed by the backend

We are going to tell Stimulus to make the controller work with the DOM of our choice. The way to do this is by means of a `data-controller` dataset:

```
<element data-controller="alias"></element>
```

In our case, we update the opening of `<main>`:

```
<main data-controller="transformer">
```

Easy, isn't it? Stimulus already has a registered controller and now knows where to apply it.

The next step is to indicate which event is related to which function, and which input is related to which target.

Managing events with actions

Actions are a structure used by Stimulus to link events to controller functions. They are declared in the DOM by means of a `data-action` dataset with the following structure:

```
<div data-controller="aliasController">
<button
  data-action=
    "event->aliasController#functionOfTheController"
>Click me!</button>
</div>
```

It will only work if it is inside a controller with the same alias; you cannot place an action in DOMs outside the tree.

Following the example, we modify our button:

```
<input
  type="button"
  value="Transform"
  data-action="click->transformer#lowercaseToUppercase"
>
```

Let's analyze what we have done with `data-action`, since it contains its own format that we must follow:

1. The event is `click`. It could be any other event, such as a `submit` event if we were in a HTML `<form>` tag, a `scroll` event, and so on.

2. After the arrow, `->`, which acts as a separator, we indicate the alias of the controller where it is enclosed.

3. Finally, after `#`, which is another separator, we indicate the function to be executed (`lowercaseToUppercase`).

We have simplified the definition of the events, but now they will also be self-managed as we include or remove elements in the DOM. Not only that, but the backend now has the superpower to add new events. Yes, you read that right, the backend can include JavaScript events! They have become datasets in the HTML that we can remove or add as needed.

There is only one step left to finish with Stimulus: detailing the inputs that can be accessed with the targets. Otherwise, we will not be able to collect the information from the forms.

Capturing references with targets

Stimulus connects to form inputs via targets, or a special dataset. Internally, Stimulus creates a variable that can be used anywhere in the controller. For example, we define in the DOM an alias called name:

```
<div data-controller="aliasController">
<input type="text" data-aliasController-target="name">
</div>
```

While in the controller, we define the following:

```
static targets = [ "name" ]
```

From here, I can call the target within any function/method in the following way:

```
this.nameTarget
```

As you can see, the alias is joined with the target text.

In the application we are developing, we have defined the target with the name myText:

```
static targets = [ "myText" ]
```

We update the DOM of the input as follows:

```
<input type="text" data-transformer-target="myText"
   placeholder="Enter text">
```

The whole frontend is ready. We have installed Stimulus, created a controller, and defined an action to trigger actions and a target to collect the input text. We only need to define the functionality in the consumer. We go to Django.

An application that converts text into uppercase letters

We have already simplified the frontend with Stimulus, installing, configuring, and implementing the tools provided by this fantastic framework. However, we still have one last step left in the application that converts the text from lowercase to uppercase: implementing the backend in the consumer.

Edit app/app_template/consumers.py with the following code:

```python
from channels.generic.websocket import JsonWebsocketConsumer
from django.template.loader import render_to_string


class ExampleConsumer(JsonWebsocketConsumer):
    def connect(self):
        """Event when client connects"""
        # Accept the connection
        self.accept()

    def receive_json(self, data_received):
        # Get the data
        data = data_received['data']
        # Depending on the action we will do one task or
        # another.
        match data_received['action']:
            case 'text in capital letters':
                self.send_uppercase(data)


    def send_uppercase(self, data):
        """Event: Send html to client"""
        self.send_json( {
                'selector': '#results',
                'html': data["text"].upper(),
        })
```

The code is so basic that it looks like it belongs in the first chapters. We have removed the actions.py file and some other elements, as we are just looking for the minimum necessary to make it work.

Let's review where the information enters the backend, where it is transformed, and where it is returned:

- The information from the frontend goes to `receive_json`, which in turn receives the `'text in capital letters'` action by executing the `self.send_uppercase(data)` function.

- `self.send_uppercase(data)` converts the text to uppercase and sends the information to the frontend, specifically to the `#results` selector.

It's time to test that everything works. We pull up Docker and go to `http://hello.localhost`. Type in the input and click on the **Transform** button.

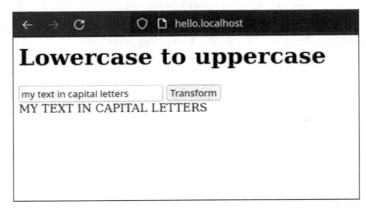

Figure 8.1 – We tested that the application works by transforming lowercase letters into uppercase

At the bottom, the text will be displayed in capitals – we've done it!

We can even improve it. The local delay between pressing the button and displaying the final result is negligible, in my case, 0.002 seconds. We can incorporate the `input` event into the input to see the result as we type, giving a feeling that there is no apparent delay:

```
<input
  type="text"
  data-transformer-target="myText"
  placeholder="Enter text"
  data-action="input->transformer#lowercaseToUppercase"
>
```

And with this small optimization, we can conclude the backend implementation of the application.

You might be tempted to take Stimulus to the examples in the previous chapters – I can only tell you: go ahead. It will be neater, you will learn a lot more about Stimulus, and things will be easier to maintain.

Summary

Our journey of learning HTML over the wire with Django comes to an end. We are now able to create SPAs in real time by gathering all the logic in the backend, avoiding duplicating tasks such as validations or HTML structures. We have relieved the frontend of a big responsibility; it now only needs to handle events or animations, thanks to Stimulus controllers and internal automation from datasets.

I would love to tell you that you already know everything you need to know, but the journey continues. The book is just a first push. There is still a lot of work ahead of you: practice, adopt Stimulus in your workflow (or any other similar framework), solve small difficulties typical of any SPA (such as managing when the user clicks on the back button in the history), explore other related protocols such as Server-Side Events, train your colleagues, convince your boss, define your line between backend and frontend (the infinite struggle of any web developer), and even adopt some other framework. The limit is set by you.

It has been a pleasure to join you in this real-time Python adventure. I have enjoyed writing every line and preparing every example. Thank you.

I can only say to you: I expect big things from you.

```
WebSocket.close()
```

Index

G

Git
  about 6
  URL 6

H

HTML
  rendering, in backend 66-72
hydration 175

I

IDE
  configuring 9-16
individual page per article
  implementing 214-216

J

JSON
  messages, sending from backend 56, 57
  used, for receiving messages
    in frontend 59-62

L

Lap 176
list of comments
  adding 217-219

M

MailHog 30, 41
matching numbers
  checking 62-66

models
  creating, for database 190-192

O

operating system 4

P

pager
  queries, limiting with 92-101
pages
  creating, with Daphne 46-48
  navigating, between with browser 210-213
  switching between 160-164
pagination
  used, for navigating between
    articles 198-203
Pillow 17
plain text
  sending, from backend 50-56
PostgreSQL 30
project
  creating 23-27
Psycopg2 17
PyCharm
  about 9
  URL 4
PyCharm Community Edition 4
PyCharm Professional 4
Python 5

Q

queries
  limiting, with pager 92-101

Packt.com

Subscribe to our online digital library for full access to over 7,000 books and videos, as well as industry leading tools to help you plan your personal development and advance your career. For more information, please visit our website.

Why subscribe?

- Spend less time learning and more time coding with practical eBooks and Videos from over 4,000 industry professionals

- Improve your learning with Skill Plans built especially for you

- Get a free eBook or video every month

- Fully searchable for easy access to vital information

- Copy and paste, print, and bookmark content

Did you know that Packt offers eBook versions of every book published, with PDF and ePub files available? You can upgrade to the eBook version at packt.com and as a print book customer, you are entitled to a discount on the eBook copy. Get in touch with us at customercare@packtpub.com for more details.

At www.packt.com, you can also read a collection of free technical articles, sign up for a range of free newsletters, and receive exclusive discounts and offers on Packt books and eBooks.

Other Books You May Enjoy

If you enjoyed this book, you may be interested in these other books by Packt:

Becoming an Enterprise Django Developer

Michael Dinder

ISBN: 978-1-80107-363-9

Use Django to develop enterprise-level apps to help scale your business.

Understand the steps and tools used to scale up a proof-of-concept project to production without going too deep into specific technologies.

Explore core Django components and how to use them in different ways to suit your app's needs.

Find out how Django allows you to build RESTful APIs.

Extract, parse, and migrate data from an old database system to a new system with Django and Python.

Web Development with Django

Chris Guest | Subhash Sundaravadivelu | Ben Shaw | Andrew Bird | Saurabh Badhwar

ISBN: 978-1-83921-250-5

- Create a new application and add models to describe your data.
- Use views and templates to control behavior and appearance.
- Implement access control through authentication and permissions.
- Develop practical web forms to add features such as file uploads.
- Develop a RESTful API and JavaScript code that communicates with it.
- Connect to a database such as PostgreSQL.

Packt is searching for authors like you

If you're interested in becoming an author for Packt, please visit `authors.packtpub.com` and apply today. We have worked with thousands of developers and tech professionals, just like you, to help them share their insight with the global tech community. You can make a general application, apply for a specific hot topic that we are recruiting an author for, or submit your own idea.

Hi!

I am Andros Fenollosa Hurtado, author of *Building SPAs with Django and HTML Over the Wire*, I really hope you enjoyed reading this book and found it useful for increasing your productivity and efficiency in Web development with Django, Python and WebSockets.

It would really help me (and other potential readers!) if you could leave a review on Amazon sharing your thoughts on this book.

Go to the link below or scan the QR code to leave your review:

`https://packt.link/r/1803240199`

Your review will help us to understand what's worked well in this book, and what could be improved upon for future editions, so it really is appreciated.

Best wishes,

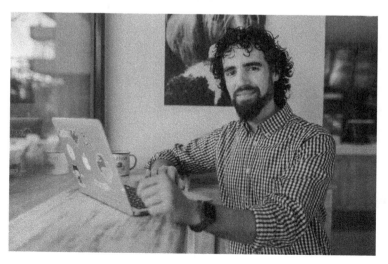